INSTITUTIONS, ECOSYSTEMS, AND SUSTAINABILITY

ECOLOGICAL ECONOMICS SERIES

Robert Costanza, *series editor*

Published and Forthcoming Titles

An Introduction to Ecological Economics
Robert Costanza, John Cumberland, Herman Daly, Robert Goodland, and Richard Norgaard

Natural Capital and Human Economic Survival, Second Edition
Thomas Prugh, Robert Costanza, John H.Cumberland, Herman E. Daly, Robert Goodland, and Richard B. Norgaard

Privileged Goods: Commoditization and Its Impact on Environment and Society
Jack P. Manno

Toward Sustainable Development: An Ecological Economics Approach
Philip Andrew Lawn

Institutions, Ecosystems, and Sustainability
Robert Costanza, Elinor Ostrom, Bobbi Low, and James Wilson

INSTITUTIONS, ECOSYSTEMS, AND SUSTAINABILITY

Edited by
Robert Costanza
Bobbi S. Low
Elinor Ostrom
James Wilson

CRC Press
Taylor & Francis Group
Boca Raton London New York

CRC Press is an imprint of the
Taylor & Francis Group, an **informa** business

CRC Press
Taylor & Francis Group
6000 Broken Sound Parkway NW, Suite 300
Boca Raton, FL 33487-2742

First issued in paperback 2019

ISBN-13: 978-1-56670-389-5 (hbk)
ISBN-13: 978-0-367-39775-3 (pbk)
Library of Congress Card Number 00-046347

Library of Congress Cataloging-in-Publication Data

Institutions, ecosystems, and sustainability / edited by Robert Constanza...[et al.]
p. cm. (Ecological economics series)
Includes bibliographical references and index.
ISBN 1-56670-389-1 (alk. paper)
1. Environmental economics—Simulation methods. 2 Sustainable development—Simulation methods. 3. Human ecology—Simulation methods. 4. Decision making—Simulation methods. I. Costanza, Robert. II. Ecological economics series (International Society for Ecological Economics).
HC79.E5 1516 2000
333.7′.01′1—dc21 00-046347

Visit the Taylor & Francis Web site at
http://www.taylorandfrancis.com

and the CRC Press Web site at
http://www.crcpress.com

Preface

Sustainability depends on understanding the way humans and their institutions interact with ecological systems. Consequently, to understand sustainability, we have to study human and natural systems together. However, scientists in the various natural and social sciences rarely, if ever, address the relevant issues together, largely because they lack a common analytic framework and language.

This book represents an attempt by scholars in diverse disciplines to cross boundaries and create a common framework that will encourage productive dialogue and synthesis across a broad range of disciplines and approaches. Creating this framework for linking human institutions and ecosystems is, we think, a prerequisite for further progress. Here we develop multiscale conceptual and mathematical models, including a range of ecosystem and human system characteristics, aimed at testing our hypothesis and providing guidance for designing sustainable human systems within sustainable ecosystems.

Section I describes our general framework (Chapter 1), some of its implications, and general principles for translating the framework to dynamic models. The framework is operationalized in dynamic models (Chapter 2). We think the major uses of the framework include:

1. Providing a common language acceptable across disciplines for developing theories and models
2. Guiding the construction of models of linked ecological and human systems
3. Organizing, synthesizing, and interpreting empirical data
4. Linking empirical data to policy processes

In Section II, we explore dynamic simulation models of diverse systems that use the framework to analyze the impact of variation in ecological (e.g., uncertainty, ecological isolation) and human (e.g., harvest rules) parameters. First, the most general version of a dynamic, linked, ecological economic model is developed (Chapter 3) and used to describe some basic behavior of these systems. Chapter 3 compares several harvest rules in the face of ecological uncertainties, spatial variation in harvest limits, and variation in the movement of resource stocks across human boundaries. Even though this model is quite simple, it exhibits many complex behaviors and subtle thresholds that highlight the complexity and difficulty of real-world resource use issues. Chapter 4 develops a more complex version of the model. This model tests harvesting rules in the context of spatial and temporal scale mismatches between the ecological system and human institutions. For example, what if, in an ocean fishery, the fish are not a large, panmictic population, but a structured metapopulation, and human rule-makers do not recognize the population structure? Chapter 5 examines a hypothetical irrigation system. It first looks at a benefit-cost analysis that would be undertaken by project designers in deciding whether to build an irrigation system. Then it examines the operation of such a system when farmers may not take the hypothetical actions posited by the planners. Chapter 6 sets these models in context, linking them to formal analytic processes and real-world problems.

Section III presents a series of studies that use models to help elucidate how resource users and policy makers view an ecological or institutional milieu, and how dynamic models can be applied empirically. Chapter 7 formalizes the mental models that lobster buyers and sellers use in their transactions, and explores how markets influence resource regulations. Chapters 8 and 9 show how dynamic models can be used as conflict-resolution devices. By making assumptions explicit (e.g., about relationships among variables), getting actors to agree on the variable values and relationships, and running the model, even management problems as complex as the Patuxent River watershed can be analyzed—and once initial conditions and relationships are agreed upon, parties in conflict are likely to acknowledge the utility of model outcomes.

In Section IV, we think about future directions and problems that might profit from the approaches we put forth here. How can we use these models to enhance and inform our decision making? As we face a new millennium, with over 6 billion people on the planet, efficient technologies and well-developed markets, the lure of short-term maximization—to take the money and run—is strong. Understanding fully the important ecological and social relationships, their interactions and temporal and spatial complexities, will require not only new techniques, but also new attitudes. Our experience suggests that perhaps these new techniques of linked models may help foster new attitudes.

About the Editors

Robert Costanza is director of the University of Maryland Institute for Ecological Economics, and a professor in the Center for Environmental Science at Solomons, and in the Biology Department at College Park. He is co-founder and past president of the International Society for Ecological Economics (ISEE) and chief editor of *Ecological Economics,* the society's journal. He currently serves on the editorial boards of eight other international academic journals. Dr. Costanza's research has focused on the interface between ecological and economic systems, particularly at larger temporal and spatial scales. This includes landscape-level spatial simulation modeling; analysis of energy and material flows through economic and ecological systems; valuation of ecosystem services, biodiversity, and natural capital; and analysis of dysfunctional incentive systems and ways to correct them. He is the author or co-author of more than 275 scientific papers (including Ecological economics: reintegrating the study of humans and nature, *Ecological Applications* 6:978–990 (1996); The value of the world's ecosystem services and natural capital, *Nature* 387:253–260 (1997); Principles for sustainable governance of the oceans, *Science* 281:198–199 (1998); and 16 books including *Ecological Economics: the Science and Management of Sustainability* (1991), *Ecosystem Health: New Goals for Environmental Management* (with Bryan Norton and Ben Haskell, 1992), *Getting Down to Earth: Practical Applications of Ecological Economics* (with Olman Segura, and Juan Martinez-Alier, 1996), *An Introduction to Ecological Economics* (with John Cumberland, Herman Daly, Robert Goodland, and Richard Norgaard, 1997) and *The Local Politics of Global Sustainability* (with Tom Prugh and Herman Daly, 2000).

Bobbi S. Low is professor and chair of the Resource Ecology and Management Concentration at the School of Natural Resources and Environment at the University of Michigan, Ann Arbor. She also holds courtesy appointments in UM's Institute for Social Research. Author of *Why Sex Matters* (Princeton) and more than 100 papers in scientific journals, she is an evolutionary biologist interested in human behavior and evolution.

Elinor Ostrom is Arthur F. Bentley Professor of Political Science and co-director of the Workshop in Political Theory and Policy Analysis and the

Center for the Study of Institutions, Population, and Environmental Change (CIPEC), Indiana University, Bloomington. She is the author of *Crafting Institutions for Self-Governing Irrigation Systems* (1992) and *Governing the Commons* (1990); co-author with Robert Keohane of *Local Commons and Global Interdependence* (1995); with Roy Gardner and James Walker of *Rules, Games, and Common-Pool Resources* (1994); and with Larry Schroeder and Susan Wynne of *Institutional Incentives and Sustainable Development* (1993).

James A. Wilson is professor of Marine Sciences and Resource Economics and director of the Marine Policy Program, School of Marine Sciences, at the University of Maine. Dr. Wilson received his Ph.D. in economics from the University of Wisconsin and has worked and taught at the Woods Hole Oceanographic Institute, the University of Western Washington, and the University of Maine. His principal writing and research interests concern the governance of human interactions with complex ecosystems, especially fisheries. He is one of a small number of resource economists who bring an interdisciplinary approach to the subject and was an early advocate of bottom-up democratic approaches to fisheries management. He has published widely, and worked and consulted with the U.S. and Swedish Academy of Sciences, a variety of federal and state agencies, and private firms and foundations.

Contributors

Roelof Boumans
Institute of Ecological Economics, Center for Environmental Science, University of Maryland

Robert Costanza
Institute of Ecological Economics, Center for Environmental Science, and Biology Department, University of Maryland

Bobbi S. Low
School of Natural Resources and Environment, University of Michigan

Thomas Maxwell
Institute of Ecological Economics, Center for Environmental Science, University of Maryland

Elinor Ostrom
Workshop in Political Theory and Policy Analysis, Indiana University

Matthias Ruth
School of Public Affairs, University of Maryland

Nirmal Sengupta
Workshop in Political Theory and Policy Analysis, Indiana University

Swati Sheladia
Workshop in Political Theory and Policy Analysis, Indiana University

Carl Simon
Center for the Study of Complex Systems, Economics Department, Mathematics Department, and School of Public Policy, University of Michigan

Ferdinando Villa
Institute of Ecological Economics, Center for Environmental Science, University of Maryland

Alexey Voinov
Institute of Ecological Economics, Center for Environmental Science, University of Maryland

Helena Voinov
Institute of Ecological Economics, Center for Environmental Science, University of Maryland

Lisa Wainger
Center for Environmental Science, University of Maryland

James A. Wilson
Resource Economics and Policy, University of Maine

Acknowledgments

The idea for this book is a product of the Social and Ecological System Linkages Project of the Property Rights and the Performance of Natural Resource Systems Workshop, The Beijer International Institute of Ecological Economics, and The Royal Swedish Academy of Sciences. Initial workshops were held on August 29–30, 1994, and May 28–June 3, 1995, in Stockholm, Sweden. An initial version of the "framework" chapter was presented at the fifth annual meeting of the International Association for the Study of Common Property, Bod, Norway, May 24–28, 1995, and we are grateful for comments and feedback from numerous individuals.

Further development took place at a series of working sessions held at the Workshop in Political Theory and Policy Analysis and the Center for the Study of Institutions, Population, and Environmental Change. We are grateful to the National Science Foundation, the Ford Foundation, the MacArthur Foundation, and to Indiana University for the support they provided for these working sessions and to university staff who worked on various aspects of this project over the long period of its gestation. Earlier versions of chapters 3, 4, 8, and 9 have appeared in journal form.

Many people helped in making the logistics work. At Indiana University we are particularly grateful to Patty Dalecki, Robert Lezotte, Anne Leinenbach, Nicole Todd, and Brenda Matthews. At the University of Michigan School of Natural Resources and Environment, Komal Bhasur put in many hours of model running and data recording. At the University of Maryland Institute for Ecological Economics, Janis King provided critical logistical support and Amanda Walker did the final technical editing of the entire manuscript.

In Chapter 7, Deirdre Gilbert of the School of Marine Sciences and Andrew Files of the Department of Resource Economics (both at the University of Maine) put in long hours helping to develop and tune the model. William Hogan and Daniel Georgianna of the Department of Economics at the University of Massachusetts (Dartmouth), and Alan Kezis, Hsiang-Tai Cheng, Stephanie Peavey, and John Holden of the Department of Resource Economics and Policy at the University of Maine prepared the secondary data, and developed and conducted the surveys and field work upon which the model is based. This work was funded by the NMFS and Sea Grant.

For Chapter 9, initial funding for this research came from the U.S. EPA Office of Policy, Planning and Evaluation (Cooperative Agreement CR821925010, Michael Brody and Mary Jo Keely, project officers). Additional funding has been provided by the U.S. EPA/NSF Water and Watersheds Program through the U.S. EPA Office of Research and Development (Grant R82-4766-010, Barbara Levinson and Brian Sidlauskas, project officers).

Our sincere thanks go to those researchers who shared their data and insights with us, including Nancy Bockstael, Jacqueline Geoghegan, Elena Irwin, Kathleen Bell, Ivar Strand, Walter Boynton, Jim Hagy, Robert Gardner, Steve Kraemer, John Jones, Rich Hall, Debbie Weller, Joe Tassone, Joe Bachman, Randolph McFarland, Patricia Steinhilber, Tom Simpson, Frank Coale, Kent Mountford, Rob Magnien, Ron Klauda, Jim Vogelman, and many others at the Maryland Department of Natural Resources, Maryland Department of the Environment and the University of Maryland Center for Environmental Science. Special thanks go to Carl Fitz for his major contributions to an earlier version of this model and continuing support.

Contents

section one

Introduction

chapter one

Ecosystems and human systems: a framework for exploring the linkages

Robert Costanza, Bobbi S. Low, Elinor Ostrom, and James A. Wilson

Contents

We humans have always had an impact on our environment. While in past times we created real environmental problems, including the elimination of numerous species, rarely have we been in as much danger as we are now, in contemporary times, of extinguishing ourselves as well as everything else on earth. When our population densities were much lower and our technologies less powerful, the stakes were not so high. Erroneous judgments based

1-56670-389-1/01/$0.00+$.50

on incomplete or faulty analysis produced bad outcomes, but not global disasters.

Once, we could muddle through and live with the consequences of our erroneous judgments. Today, there are more actors in the world—more divergent interests, more interactions, and an accelerating pace of change. Fortunately, there has been growth in our analytic and computational tools. Only 20 years ago, it took a room full of computational machines to model the U.S. economy—a problem that today takes a few seconds on an inexpensive laptop. We have made great strides in reducing complex real-world problems to models that, while simpler than the real world, retain crucial aspects of the problems under analysis, and can be used for analysis and experimentation. We now can take into account a very large number of variables; we can model complex, non-linear interactions in a way that was impossible just a few years ago. In addition, we can also develop and use models of the complex interactions between ecological and human systems. In many cases, we can model our way to solutions rather than just muddle through.

While models of complex ecological and human systems are essential tools that can help resource users, public officials, and stakeholder groups to develop more effective environmental understandings and policies, models themselves must be understood for what they are. Models are always a simplification of a complex world. Further, models can only be as precise as the data that are used to calibrate them—and a surprising proportion of models are not calibrated. Thus, it is essential for scientists and stakeholders to view models as useful tools but not as infallible guides to future action somehow representing Truth.

We are proposing something quite radical: connecting ecosystems models (a relatively well-developed field within ecology) with models of human systems (a less well-developed field). These two fields traditionally approach analysis in significantly different ways; they have different languages and different foci. Few scholars in either area have communicated actively with scholars in the other; however, it is this combination of disciplines that is necessary to fully analyze today's environmental problems. Studying the relevant ecological aspects of the world, relevant cultural phenomena, and the relationships among rule systems and among particular rules requires multiple disciplines, multiple disciplinary languages, and multiple levels of analysis.

1.1 Frameworks, theories, and models

To study the human decision making that affects ecological processes requires theoretical work undertaken at three levels of specificity: (1) frameworks, (2) theories, and (3) models. These levels are often confused with one another but recognizing their distinctness is important because analyses conducted at each level provide different degrees of specificity related to a particular set of problems.

The development and use of a general framework helps to identify the elements and the relationships among these elements that one needs to consider in both institutional and ecological analyses. Frameworks attempt to identify the universal elements of any theory (the essential elements to include in an analysis). They provide a metatheoretic language for comparing theories, something particularly important when we attempt interdisciplinary work. Frameworks organize diagnostic and prescriptive inquiry, and provide the most general list of variables that should be used in analysis. The elements contained in a framework help the analyst identify the central questions that need to be addressed.

Theories enable the analyst to specify which elements of the framework are particularly relevant for certain kinds of questions, and to make general working assumptions about these elements. Thus, within a framework theories make specific assumptions that are necessary for an analyst to diagnose a phenomenon, explain its processes, and predict outcomes. Typically, several theories are compatible with any one framework; it can be difficult to determine true alternatives. Economic theory, game theory, transaction cost theory, social choice theory, covenantal theory, and theories of public goods and common-pool resources are all compatible with the framework discussed in this book.

Developing and using models requires explicit and precise assumptions about a limited set of parameters and variables. Logic, mathematics, game theory, experimentation, simulation, and other means are used to explore systematically the consequences of these assumptions on a limited set of outcomes. Most theories are compatible with multiple models. For example, when Weissing and Ostrom (1991, 1993) sought to understand the strategic structure of the games that irrigators play in differently organized irrigation systems—how successful farmer organizations arranged for monitoring and sanctioning activities—they developed four families of models within a single theoretical structure to explore the likely consequences of different institutional and physical structures.

1.2 Sustainability

Sustainability broadly refers to the persistence of the integrity and structure of any system over time; the concept is thus of central interest to both ecologists and policy analysts who study resource use. There is considerable debate about how to define sustainability, sustainable development, and related concepts (cf. Pezzey, 1989; World Commission on Environment and Development, 1987; Costanza, 1991; Pearce and Atkinson, 1993).

Critics argue that the concept of sustainability is useless because it cannot be adequately defined. Much of this discussion is misdirected because critics (1) fail to take into account the range of time and space scales over which the concept must apply; and (2) fail to realize that the real problems are related to prediction rather than definition. Costanza and Patten (1995) suggest a relatively robust definition that will serve most uses:

> A sustainable system is a renewable system that survives for some specified (non-infinite) time.

Biologically, this means the resource avoids extinction. Economically, it means human resource users avoid major disruptions and collapses, and can hedge against instabilities and discontinuities. Sustainability, at its base, always concerns time, and, in particular, longevity. The difficulty with this definition is that, like some definitions of "fitness" in evolutionary biology (Dawkins, 1982, Chapter 5), determinations of relative success sometimes can only be made after the fact. An organism alive right now is (informally) "fit" to the extent that its progeny survive and contribute to future generations, but calculating fitness must wait until tomorrow when relative survival and reproduction are known. The assessment of relative sustainability must also wait for the future, but we have a stake in biasing the outcome.

What often pass as definitions of sustainability are frequently just predictions that particular actions taken today will lead to sustainability. For example, keeping harvest rates of a resource system below rates of natural renewal should, one can argue, lead to a sustainable extraction system—but that is a prediction, not a definition. Prediction is, in fact, the foundation of maximum sustainable yield (MSY) theory, which was for many years the basis for managing exploited wildlife and fisheries populations (see Chapters 3 and 4). The sustainability of a system can only be known after sufficient time has passed to observe whether the prediction held true. So much uncertainty exists in estimating natural rates of renewal, and observing and regulating harvest rates, that a simple prediction (i.e., MSY) is always highly suspect, especially if it is erroneously thought of as a definition (Ludwig et al., 1993). This will be a recurring theme in later chapters.

Rapid feedback and appropriate selection mechanisms can sometimes compensate for lack of knowledge by decision makers. If over harvesting immediately produces local and visible resource reduction, lower harvesting likely follows as a matter of course. However, in the interactions between ecosystems and human systems such processes are often slow, and feedback may come too late.

Despite the well-documented difficulties of designing sustainable resource systems, there are well-documented examples, past and present, of natural resource systems that have proven effective and sustainable over time (Gibson et al., 2000; Ostrom, 1990; Berkes, 1989; Bromley et al., 1992; Lam, 1998). Many of these have evolved over long periods of time as human actors have learned more about how a local ecosystem reacts to various harvesting and investment strategies. These regulatory systems frequently appear complex and nonsensical to external observers. Efforts to devise simple regulatory policies for large areas have often threatened the sustainability of both natural resources and previously effective governance systems (Atran, 1993; McCay and Acheson, 1987; Wilson, 1990).

We hypothesize that the causes of many sustainability problems lie in "scale" problems. Large-scale ecosystems are not simply small-scale systems grown large, nor are micro-scale ecosystems mere microcosms of large-scale systems. The driving forces and feedback mechanisms in large- and small-scale systems operate at different levels and exhibit distinct patterns. This means that management systems that produce acceptable outcomes when applied to ecosystems at one level can (and frequently do) produce disruptive or destructive results when applied to higher level or lower level systems. Management practices that do well in handling traditional resource uses at the local level, for example, cannot be expected to do equally well in handling activities organized at a continental or global scale. Even more important, when local systems are fully superseded by national or international management practices, local ecosystems frequently suffer (Finlayson and McCay, 1998; Arnold, 1998). The solution, then, is to match ecosystems and governance systems in order to maximize the compatibility between these two types of systems.

1.3 *Hierarchy and scale problems*

In modeling complex systems, scale and hierarchy are central issues (O'Neill et al., 1989). In some sense, the natural world (including humans) contains a convenient hierarchy of scales based on interaction-minimizing boundaries: scales ranging from atoms to molecules to cells to organs to organisms to populations to communities to ecosystems (including economic and human-dominated ecosystems) to bioregions to the global system and beyond (e.g., Allan and Starr, 1982; O'Neill et al., 1986). "Scale" in this context refers to both the resolution (spatial, temporal, or degree of complication) and extent (in time, space, and number of components modeled) of the analysis.

Multi-scale phenomena are particularly prevalent in many natural resource systems and many human institutions. We argue that analytic failure to recognize this fact has led to persistent problems. In both ecology and economics, primary information and measurements are generally collected on relatively small scales (e.g., small plots in ecology, individuals or single firms in economics); oftentimes, that information is subsequently used to build models at radically different scales (e.g., regional, national, or global). If we are correct that large systems are not "small systems grown large," this process is directly tied to the problem of aggregation (the process of adding or otherwise combining components). In complex, non-linear, discontinuous systems—like ecological and economic systems—aggregation is a far-from-trivial problem (O'Neill and Rust, 1979; Rastetter et al., 1992). For example, in applied economics, basic data sets are often derived from national accounts, which contain data that are linearly aggregated over individuals, companies, or organizations. Sonnenschein (1974) and Debreu (1974) showed that, unless one makes very strong and unrealistic assumptions about the individual units, the aggregate (large scale) relations between variables have no resemblance to the corresponding relations on the smaller scale. Serious work to develop multi-scale models of either ecosystems or human institutional systems has only begun very recently.

If ecosystems actually functioned as a seamless web with no practical subdivisions, understanding and managing such systems would require a massive, centralized modeling, measurement, and monitoring effort. Any missing piece or assumption could render the model useless. On the other hand, if ecosystems can be partitioned into relatively separable parts that are largely understandable on their own, measurement and monitoring requirements might still be great, but the ability to partition the problem would make understanding and managing the overall system much more tractable. More importantly, the need to pass large amounts of information along to a centralized management structure could be reduced greatly with little loss of understanding or management capability. Finally, the variety and appropriateness of regulatory responses could be increased if each subunit responded independently to local disturbances; this would enhance the overall responsiveness of the system (Ashby, 1960).

Fortunately, most ecosystems do appear to function as partitionable systems. For example, Allen and his colleagues (1982) suggested that for all practical purposes ecosystems function as connected subsystems. Each subsystem existing in a certain place for a certain time of the year (e.g., a nursery ground) can be treated as a separate entity during that period. When this is the case, the scientific and practical understanding of the system can be divided into tractable smaller problems. Just as important, the day-to-day management of a large number of ecosystem functions can be assigned to local management units with no need to pass information about the condition or state of a particular subsystem on to more central and removed management authorities. However, if there are no human institutions at a small level, it is impossible to effectively utilize information about diversity and local variation.

Nevertheless, most subsystems, both human and non-human, cannot be managed with complete independence; they are connected to the rest of the system. Events within a subsystem can be treated as independent up to a point, but such independence is constrained by the fact that the larger system relies on contributions of the various subsystems, and the subsystem relies on contributions from other subunits. Events within a subsystem can affect the rest of the system through phenomena such as migration, runoff dispersal, temperature changes, and so on. If there are no human institutions at the level of these larger systems, it is impossible to regulate these and other transboundary phenomena.

The Everglades of South Florida provide an interesting example of how partitioning our analyses and recognizing multi-level phenomena could improve our analytic ability. The agricultural areas south of Lake Okeechobee have been suggested as possible sites for combined wet agriculture (using wet adapted strains of sugar cane and other crops) and water storage (which tends to reduce or stop the subsidence of soils and diminish the phosphorus load of waters entering the sawgrass regime). Management of this "subsystem" might proceed almost independently of the rest of the ecosystem so long as it operated under the constraint of having to release water to lower parts of the system in appropriate seasonal flows. A market-like arrangement, in which water managers used variable monthly water storage lease prices, could work effectively to signal times when water was required in the lower part of the system. That is, it could integrate the functioning of the agricultural areas with the rest of the ecosystem.[1] Managers working with system-wide hydrology would need to know only how much water was being held. They could alter that amount by varying the prices they were willing to pay to lease storage. Individual field management would remain with individual farmers—those most knowledgeable about the relevant circumstances at that scale. The Maine lobster fisheries (Chapters 3, 4) are another clear example of how recognizing multiple scales of interaction can solve a mismatch of rules and ecological realities.

1.4 *Uncertainty, limited information, and misplaced certainty*

To understand the scope of any problem, we must determine what we do, and do not, truly know. We must differentiate between risk (an event with a known probability, sometimes referred to as statistical uncertainty) and true uncertainty (an event with an unknown probability, sometimes referred to as indeterminacy). Many important environmental problems suffer from true uncertainty, not merely risk. The scientific method treats uncertainty as a given, a characteristic of all information that must be honestly acknowledged and communicated. Over the years, scientists have developed increasingly sophisticated methods to measure and communicate uncertainty arising from various causes. In general, the progress of science tends to uncover uncertainty more often than it reveals absolute results. However, the lay public tends to mistakenly interpret "scientific" results as absolutely precise. Policy makers would also like to assume that scientific knowledge can eliminate uncertainty, but the scientific method can only set boundaries on the limits of our knowledge. Ecological analyses can tell us the range of uncertainty about global warming, the potential impacts of toxic chemicals, or the possible range of fish population dynamics, and maybe something about the

[1] A leasing system of this sort would effectively alter the costs of production of wet and dry crops, favoring wet crops. This same change in relative costs would also encourage experimentation with new and/or untried wet crops and, if supported by appropriate plant breeding programs, would provide the incentives for the gradual transformation of agriculture in the area.

relative probabilities of different outcomes. In most important cases, however, such analyses cannot tell us which of the possible outcomes will occur with any degree of accuracy.

Environmental managers and policy makers, on the other hand, would prefer confident answers from ecologists, for clear and cogent reasons. The goal of policy is making unambiguous, defensible decisions, often codified in the form of laws and regulations. Legislative language is often open to interpretation, and regulations are easier to write and enforce if they are stated in clear and absolutely "certain" terms.

Policy-makers of most contemporary environmental regulations, particularly in the United States, demand certainty. When scientists are pressured to supply this nonexistent commodity, the result is frustration, poor communication, and mixed messages in the media. Uncertainty exists, but its existence means that environmental issues can be manipulated by political and economic interest groups. Uncertainty about global warming is perhaps the most visible current example of this effect. If we hope to use scientific analyses to make policy, we need to deal with the whole array of possible futures and all their implications, and not delude ourselves that certainty is possible. We need to stress the importance of developing institutions that tend to generate accurate information about ecosystem structure and use so that the edge of uncertainty is slowly pushed back over time.

1.5 *Conflicts of interest*

Conflict over resources is universal among living organisms; in a finite world, resources are a zero-sum game. Genetic conflicts of interest exist among all living organisms: individuals strive to increase their own inclusive fitness (Hamilton, 1964; Grafen, 1991: 9–13) at the expense of non-related individuals in a finite world. Cooperation, in both human and non-human systems, is likely under specific conditions: the number of actors is smaller rather than larger, interactions are repeated, and actors are able to detect cheating and punish offenders. Both the ecological and the social science literature are converging on such findings (e.g., Ostrom, 1990; Ostrom et al., 1999; Keohane and Ostrom, 1995; Alexander, 1987; Low, 2000; Dugatkin, 1997).

For simplicity, management initiatives may assume a unified bureaucratic actor and a unified community of resource users—but conflicts are common among and within units at every scale. At the smallest unit of analysis, even within households, men and women may have different, and occasionally conflicting rather than complementary, resource use systems (Low, 1994; Carney and Watts, 1990). For example, men who invest their wealth in livestock may come into conflict with women concerned with cultivation of the same land. At higher scales, stratification of communities on the basis of religion, occupation, wealth and class, ethnicity, and longevity of residence can result in conflicting rather than complementary resource use (Cernea, 1985; Nhira, 1994; Briscoe, 1979; Brown, 1995).

At even higher scales, consider conflicts among different communities trying to use the same resource: conflicts over water use in arid zones or over fishing in both coastal waters and the high seas (Berkes and Folke, 1998). For decades, scholars of bureaucracy have documented fragmentation, parochialism, and conflict, both within and among units of bureaucracies in societies of all types. Different units of a government's bureaucracy may work smoothly enough with their own interest group clients—say, the U.S. Forest Service with the timber industry and the U.S. Park Service with environmentalists—but will have conflicts over how best (and how much) to use a resource. These within- and among-unit conflicts parallel the ecological examples above. The obvious implications for successful analysis are that (1) resource management systems must be examined at different scales, and (2) the activities, interests, and outcomes for different categories of actors in units at the same scale must be differentiated.

1.6 Sustainable ecosystems and human systems

Underlying causes for the mismanagement of natural resources tend to be associated first with missing or failed institutions, and second, with scale mismatches among institutions:

1. *Missing Institutions:* human institutions do not exist at the appropriate scale or have not established effective controls of ecosystem stocks and flows. This typically results in open access systems and resource degradation.
2. *Scale Mismatches:* potentially effective institutions exist at the appropriate scales, but the following must be considered:

   A. *Missing Connections:* decision making linkages between scales are ineffective

   B. *Incorrect Scale of Information:* decisions are based on information aggregated at the wrong scale, even though information may exist at the appropriate scale.

Problems of missing or mismatched institutions can arise because human systems of rules and mechanisms for coordination and control are rooted in history and reflect past and present struggles over the distribution of wealth. This path-dependency means that property rights have been divided and partitioned in ways that may not correspond to the scale and structure of ecosystems.

Consider migratory species of fish in the ocean; they typically travel from the jurisdiction of one national state to another. Systematic use of these stocks requires cooperation and contracting between sovereign states, which is often problematic because mechanisms for credible commitments and effective enforcement are lacking. For instance, if the fishers in state A know that particular fish stocks are about to leave their fisheries zone for the jurisdiction of state B, the fishers (and their government) have a strong incentive to decimate the stocks before they migrate. When valuable fish are found in waters outside all national fisheries zones, contracting for the rational use of these stocks is clearly problematic. Even when current users succeed in constraining their own behavior, the agreement can act as an incentive for outsiders to enter these waters. Within national states, property rights are partitioned between internal political jurisdictions (for instance, between central and regional authorities), which can create difficulties comparable to those found in international relations. However, states usually have better means than the international community to solve their internal conflicts.

All these problems share a common feature. Because control does not match the scale and structure of the ecosystem, the actors who consider their individual costs and benefits fail to consider the full costs and benefits of their actions, including those costs and benefits that fall on others. The remedy involves structural changes to reduce these external effects, but political considerations may be intractable.

Poor design of institutions and incentives can also promote resource depletion. A polity may have full jurisdiction over a particular ecosystem and still introduce a system of rules, coordination, and enforcement that fails to sustain ecosystems. Recently created biosphere reserves in many countries of the world assign full control over all resources in the reserve to a national government, but find that resource deterioration continues. This situation is a variant of the missing institutions definition above. There are various reasons for this failure:

1. Polity leaders and members may perceive the costs of establishing and enforcing an effective system of property rights as greater than the benefits (Eggertsson, 1990). In other words, the expected opportunity cost is greater than the community is willing to bear. Only events that change the costs or benefits of establishing and operating an effective system will lead to changes in behavior.
2. The political system for aggregating preferences or for collective action may produce outcomes that are not consistent with the individual preferences of the actors involved (e.g., systems of representation, voting issues; Arrow, 1951).
3. Independent of costly measurement and enforcement or perverse political outcomes, actors may simply not desire to bear the opportunity cost of sustaining a particular ecosystem. Changes in behavior will require a new perception of costs and benefits.[2]

[2]Particularly complex situations may arise when local users of a resource are asked by outsiders to bear all the cost of reducing the rate of utilization: consider the Spotted Owl debate. Sometimes, outsiders may be willing to share in the cost of reduced utilization, but transaction costs and problems with jurisdictions prevent contracting between insiders and outsiders. Finally, the outsiders may be ready to use force (trade sanctions or even violence) to change the users' behavior.

Mismatched systems can go awry in at least two ways. First, current governance or management systems may be unconnected to other parts of the human system at larger or smaller scales (2A above). How do we solve a real-world problem by constructing a model? Second, decision makers may rely on ecological information aggregated at too small or too large a scale (2B above). For example, in the cod fisheries of eastern Canada, local fishermen reported a decline in the size of the fish that they were capturing, reflecting lowered recruitment (Finlayson and McCay, 1998). However, catch data considered by governmental officials concerned tonnage, not size, and were aggregated to include both off-shore and in-shore information. The local information was swamped by the aggregation process, and the fisheries decline became serious before any management agency realized what was happening (see also Chapter 4).

Knowledge about the nature and structure of systems can be critically important. In the 1960s and 1970s, textbooks in economics frequently claimed that centrally managed societies would be more likely to protect the environment than decentralized market economies because the central managers would internalize all relevant costs and benefits (e.g., Ophuls, 1973; Heilbruner, 1974). Similarly, central control was posited as the solution to commons problems (Hardin, 1968). As we have gained knowledge, we can see that these assertions are not necessarily true (e.g., Ostrom, 1990; Ostrom et al., 1994). Equally naive assertions that local resource control is always best have similarly proved unfounded; how well local control works depends on the perceived costs and benefits of local actors, their dependence on the system over the long term, their nesting in a larger system, and the presence of adequate conflict-resolution mechanisms (Ostrom, 2000).

1.7 Human–ecosystem relationships: a framework

Ecologists and social scientists use different languages, frameworks, theories, and models. It is no wonder they have difficulty understanding each other's worlds and how those worlds interact.[3] To help us understand and model the relationships between ecosystems and human systems, we need a common language and an adequate conceptual framework within which to work. The lack of this framework has hindered communication between the relevant disciplines and has limited progress on sharing data, concepts, models, and results.

Developing a lexicon of key terms is an essential part of our task because many important concepts used in the relevant disciplines are not known and understood in other disciplines. If there is to be a serious joint effort by

---

[3]Even without such communication problems, achieving agreement on the meaning of empirical tests of theories is difficult, because data collected in field settings do not necessarily include measures of all relevant variables (particularly those variables that are considered external to the system, and are frequently stochastic and unmeasured). Thus, field studies of ecosystems may of necessity ignore important economic, social, and political variables; and studies of human systems frequently ignore important ecological variables.

scholars from diverse disciplines that goes beyond working on different parts of a large project; those participating must begin using concepts and terms in a similar way.

Here we begin development of a common framework, and suggest how it can be used. Figure 1.1 represents the parallels between human and ecological systems, and the nature of their interactions. Both ecological and social systems have "stocks," "flows" among those stocks, and "controls" of those flows. The stocks, flows, and their interactions have similar attributes in both systems, even though particulars differ. In an ecosystem, the biomass of fish, for example, comprises one stock. Stock can flow from a fish population into a fisheries catch; this flow can be predictable or unpredictable.

The human systems and ecosystems interact, and all such interactions also have flows, controls, and attributes (Figure 1.1). By structuring the human and ecological systems and their interactions in parallel forms, we hope to facilitate comparisons at multiple levels.

1.7.1 Stocks

By stocks we refer to elements in the system that can potentially accumulate or decline. Here we include both capital and actors.

Human-made capital (assets) constitute the material and non-material resources actors use to pursue activities. Many scholars differentiate among three types of human-made capital: physical, human, and social. Although money is sometimes thought of as a form of capital, it is really just a means of exchange and a way to store value. Consequently, it can be used to purchase any of the kinds of capital we discuss here. Physical capital comprises the factories, buildings, tools, and other physical artifacts usually associated

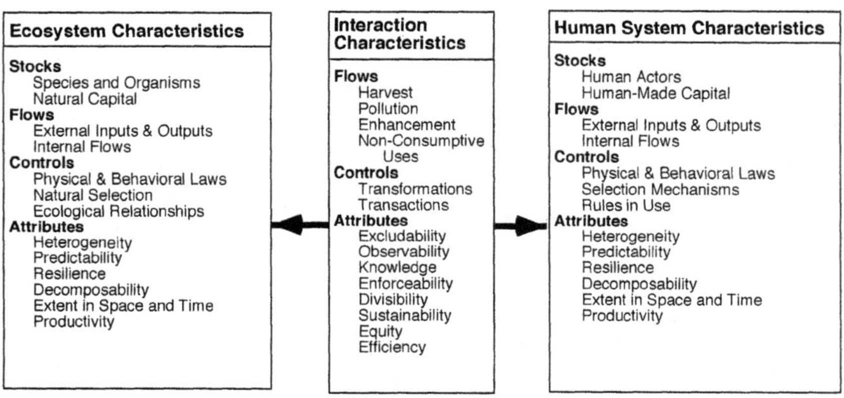

Figure 1.1 A framework for analyzing human and ecosystem interactions. Note the parallel entities and processes in both ecological and human systems.

with the term "capital." This form of capital is inactive unless people activate it. Human capital is the stock of education, skills, culture, and knowledge stored in individual human beings. A third sort of human-made capital that has been discussed and analyzed more recently is social capital (Coleman, 1988), the commonly shared and understood relationships that can enhance mutually beneficial outcomes of a process. Social capital involves regularized patterns of linking actors in patterned sets of relationships and must be shared among individuals rather than possessed by a single individual. Social capital is not captured by the traditional factors. Families, clans, firms, and governments are all examples of social capital. Each type of organizational structure has distinctive patterns of stratification, dominance, capabilities, and limits. Any particular organizational structure will be characterized by the number of its members, its geographic and temporal extent, its history, the assets it controls, its information generating and processing capability, its production technology, and so forth.

Because capital is traditionally defined as something produced or manufactured that is also a means of production of other goods, the term "natural capital" needs explanation. It is based on a functional definition of capital as "a stock that yields a flow of valuable goods or services into the future." What is functionally important is that a stock yields a flow—whether the stock is manufactured or natural is, in this view, a distinction between kinds of capital and not a defining characteristic of capital itself. Natural capital may also provide services like recycling waste materials or water catchment and erosion control, which are counted as natural income.

We differentiate two broad types of natural capital: renewable and non-renewable. Renewable natural capital is active and uses solar energy to be self-maintaining; it can be harvested to yield ecosystem goods like wood. When left in place, renewable natural capital yields a flow of ecosystem services like erosion control and recreation (Costanza et al., 1997). Non-renewable natural capital (e.g., fossil fuel and mineral deposits) generally yields no services until extracted. Further, non-renewable natural capital like fossil fuel exists in hidden deposits of unknown size—which can create perverse economic incentives (Low and Heinen, 1993).[4] Renewable natural capital is analogous to machines and is subject to entropic depreciation; nonrenewable natural capital is analogous to inventories and is subject to liquidation (El Serafy, 1989). Thus we have three types of human-made capital (physical, human, and social) in addition to the natural capital discussed

[4]When harvesting or extraction gives rapid feedback (e.g., over harvesting of a local fish), harvest adjustments are likely; this is typically the case for renewable resources. In contrast, when the extent of the resource is unknown (more common among non-renewable resources), the rational strategy is to take what is possible. Thus, Pleistocene water resources, oil and gas reserves, and open-ocean whaling are difficult to manage over the long term with conservation as the basis for management. For example, Clark (1976) noted that the economically rational management of whales was to harvest them all and bank the money, for money grows faster than whales.

above. Natural, human, and manufactured capital correspond roughly to the traditional economic factors of production of land, labor, and capital.[5]

In addition to these forms of capital, the stocks shown in Figure 1.1 include organisms and species. In human systems, individuals (and aggregations of individuals) are the human actors, individuals who, like natural capital, comprise a stock. Individual organisms are actors in ecological systems; they interact not only with conspecifics (as in human political and social interactions) but also with individuals of other species (e.g., in predator-prey and competitive interactions). As a result, species persist, increase or decrease over time, and shift in geographical distribution. Humans are more complicated: they make choices among actions leading to different outcomes, and can have great impact on non-human portions of the system.

1.7.2 Flows

Flows (Figure 1.1) are the transactions or exchanges of material assets or information from one stock to another in all human systems and ecosystems. External inputs and outputs of energy are universal in ecosystems. Internal transfers of energy, and flows of matter, vary in scale and speed across ecosystems. Failure to recognize these rate differences is, as we note above, one source of problems of "scale mismatch." For example, a stock or population of trees or fish provides a flow or annual yield of new trees or fish, a flow that can be sustainable year after year. The sustainable flow is natural income; the stock that yields the sustainable flow is natural capital.[6] In addition, stocks such as fish may move among ecosystems (measured as the natural transfer rate in the models of Section II).

Humans value particular ecological flows for consumption (e.g., volume or biomass of particular species of fish, or of lumber from specified tree species), giving rise to human-ecosystem interactions (see "Interaction Characteristics" in Figure 1.1). For example, these flows can be harvested through the removal of natural material from the ecosystem. Other outputs arising from the interactions of the human and ecological systems include (1) pollution, or the return of waste products from transformations of material and energy to the ecosystem; (2) enhancement, or investment in maintenance or restoration of resource quality and resource productivity; and (3) non-consumptive uses such as certain kinds of recreation (see Costanza et al., 1997).

---

[5]These actors are also a stock, parallel to organisms in ecosystems—but they can affect other parts of systems in special ways. The expectations and values that actors associate with activities affect their actions and outcomes. Individuals hold perceptions (information, beliefs, and models) about causal processes and about the state of particular variables. Individuals' models differ in their degree of completeness, accuracy, fineness, and complexity, and in how much information is retained and integrated in memory.

[6]Since the flow of services from an ecosystem requires that the whole system be functional, the structure and heterogeneity of the system are important attributes of natural capital (Figure 1.1; also see Costanza et al., 1997).

1.7.3 Controls

All systems involve controls (Figure 1.1). In ecosystems, physical and behavioral laws control many processes (e.g., temperature controls the speed at which many reactions can occur). Natural selection, the rules governing the survival and reproduction of all living things, interacts with physical laws to constrain the life histories and behavior of living components of ecosystems. For example, in consistently cold regions of the Arctic, we know that the survival and reproduction of fish species has been shaped over time by consistently cold conditions, exemplifying the differential success of cold-tolerant species versus other individuals. Ecological relationships (e.g., competition, predator-prey, mutualism) result from the interaction of physical laws and natural selection, and further constrain the type and complexity of interactions that can occur in ecosystems.

In human systems, controls include physical and behavioral laws, selection mechanisms, and rules in use. Behavioral laws include determinant or probabilistic responses to stimulus—"knee-jerk" responses, limits on human attention and cognition, and psychological responses to "charismatic" traits. Human selection mechanisms choose some individuals or organizations from an available population to be rewarded or punished, and thus increase or decrease their likelihood of persistence. Examples include entrance examinations for college, job requirements, and profitability in economic systems. Rules in-use are enforceable constraints on actions and outcomes placed by humans on themselves and others. These rules exist at multiple levels, and always in the context of the community in which they are jointly understood and enforced. Rules define the actions that individuals may, must, or must not take (our definition here is not equivalent to formal laws, which are formulations made by legislatures, executives and administrative agencies, and courts) (Crawford and Ostrom, 1995). When we refer to a human system as one governed by the rule of law, we usually mean that there is a close correspondence between rules-in-form (*de jure*) and rules-in-use (*de facto*).[7] The following (taken from Chapter 2 of Ostrom et al., 1994) are seven key types of rules that affect the structure of organizational arrangements:

1. *Position rules* specify a set of *positions* and how many participants are to hold each position. Example: Farmers who constitute an irrigation association designate positions such as member, water distributor, guard, member of a tribunal (to adjudicate disputes over water allocation), and other officers of the association.
2. *Boundary rules* specify how *participants* enter or leave these positions. Example: An irrigation association has rules that specify how a farmer becomes a member of the association and the qualifications that individuals must have to be considered eligible to hold a position as an officer of the association.
3. *Authority rules* specify which *set of actions* is assigned to which position at each node of a decision tree. Example: If a farmer challenges the

actions taken by another farmer or the water distributor, the rules of an irrigation association specify what a water distributor or guard may do next.

4. *Aggregation rules* specify the *transformation function* to be used at a particular node, to map actions into intermediate or final outcomes. Example: When a decision is made at a meeting of an irrigation association about changing association rules, the votes of each member present and voting are weighted (frequently each vote is given equal weight, but it may be weighted by the amount of land owned or other factors) and added. When 50 percent plus one of those voting (presuming a quorum) vote to alter legislation, the rules are altered. If less than 50 percent plus one vote for the change, the rules remain unchanged.
5. *Scope rules* specify the *set of outcomes* that may be affected, including whether outcomes are intermediate or final. Example: Rules that specify that the water stored behind a reservoir may not be released for irrigation if the level falls below the level required for navigation or for generating power.
6. *Information rules* specify the *information* available to each position at a decision node. Example: Rules that specify that the financial records of an irrigation association must be available to the members at the time of the annual meeting.
7. *Payoff rules* specify how *benefits* and *costs* are required, permitted, or forbidden in relation to players, based on the full set of actions taken and outcomes reached. Example: Rules that specify whether a farmer may sell any of the water received from an irrigation system, what crops may be grown, how guards are to be paid, and what labor obligations may be involved to keep the system maintained.

In the interactions between humans and ecosystems, two controls are of central importance, transformation and transactions. Transformations are physical changes of inputs into outputs; production and consumption represent two major transformations when humans interact with ecosystems. Transactions are the transfers from one party to another in exchange relationships (rights to inputs, outputs, and assets). For example, when someone harvests timber or produces paper, these are transformations; when they hire workers or sell timber or paper, these are transactions.

1.7.4 Attributes

Attributes are the characteristics of stocks, flows, controls, and their relationships (Figure 1.1). The number of attributes that potentially affect the capacity of human actors to manage resources sustainably is very large. We concentrate here on a limited number of attributes to capture important variation in functionally significant ways: heterogeneity, predictability, resilience, decomposability, extent in space and time, and productivity.

Heterogeneity reflects variation in the attributes of entities. Heterogeneity is low when most human or ecosystem entities are similar in structure and value (homogeneous), and high when many entities differ in structure and/or value. In human systems, individuals vary along important continua, including age, wealth, skill, and strength. Such differences clearly can affect demands on an ecosystem, conflict among individuals and groups, and the challenge of crafting institutions and incentives to increase sustainability. Predictability measures the degree to which any entity's behavior can be forecast as a measure of the degree to which it remains constant, or, if it changes, the degree to which those changes can be predicted as a function of some other variable or entity. (For example, are the fluctuations seasonally cyclic?) Resilience measures the magnitude of disturbances that can be absorbed before a system shifts from one locally stable equilibrium to another (Holling, 1987). For example, if there is unsustainable harvesting of a commercial fish species, "trash fish" may increase in numbers and replace the commercial species. Resilience measures the resistance of systems to such shifts. Decomposability measures the degree to which distinct components of the system can be broken down into discrete, small, homogeneous aspects. Most complex systems are organized into nearly decomposable (i.e., independent) subsystems. Subsystems are typically connected to one another through horizontal mechanisms at the same scale and, at the same time, contained within nested, vertical hierarchies. Extent in space and time reflects the size of the geographic region covered by stocks of ecological or human systems or the length of time considered. The terms "patchiness" and "grain" are used to reflect scale in terms of area; longevity reflects a temporal scale. Productivity characterizes the amount of outflow obtainable from a particular combination of stocks, inflows, and controls.

1.7.5 Interactions

All of the above attributes can be measured in both human systems and ecosystems. When human systems and ecosystems interact (Figure 1.1) an additional set of attributes is important that characterizes the relationships between human systems and ecosystems: excludability, observability, knowledge, enforceability, divisibility, and sustainability. Excludability refers to the capability and cost of keeping some individuals from benefiting from a system. Observability is the capability of detecting and measuring human actions and their consequences on ecosystems and human systems. Knowledge represents the level of understanding of how the system is structured and the relevant values of key variables by those using a particular resource system. The level of knowledge about a particular system may vary from full certainty to considerable uncertainty. Enforceability reflects the feasibility and cost of achieving conformance to rules. Divisibility refers to the separability of a resource into units that can be used by different individuals. This attribute is frequently referred to as subtractability of the flow of benefits or the rivalry for the benefits produced. Sustainability, defined above because it is so central to all management issues, reflects the persistence of a stock over time as it is used.

In any human-ecosystem interaction, the costs and benefits of these attributes are centrally important. Consider fisheries: sustainability will vary with the ecology of the particular fish species (e.g., as a function of reproductive or recruitment rate). If harvesting and investment decision makers have relatively complete and certain information about the structure of the ecosystem, good information about the values of key variables, and long-term interests tied to the resource, the likelihood of achieving long-term sustainability is enhanced. If it is impossible or too costly to exclude outsiders, free riders can destroy the system even if no other problems exist. If one cannot observe fishers on the fishing grounds, or if they do not return to the same harbor, it may be difficult to monitor the equipment they use or the amount they catch and thus difficult to enforce catch rules. Any of these conditions decreases the possibility of achieving sustainability.[8]

1.8 Complex systems and dynamic models

We are interested in studying complex, nested systems in which it is extremely difficult for any actor to obtain appropriate knowledge about the nature and structure of ecosystems and humans systems and thus, their dynamic behavior over time. Here we have argued that one basic pitfall involves a mismatch between the structure and attributes of ecosystems and the structure and attributes of connected human systems. To explore these mismatches, however, requires that ecologists and social scientists develop a common language and framework for studying complex, hierarchical systems that involve considerable uncertainty. To this end, we propose a coherent framework that can be used by both ecologists and social scientists when modeling these kinds of coupled systems over time.

The modeling and empirical studies in the remainder of this book apply our framework to some of the policy issues raised above—particularly the effects of diverse human systems on the sustainability of complex ecosystems. In Chapter 2, we discuss ways to make use of the framework as models able to incorporate empirical data. Section II presents a group of simulation models based on the general framework. The models make different tradeoffs among the characteristics of generality, realism, and precision. In Section III, we explore ways to use such models in the real world in teaching, hypothesis testing, and conflict resolution. Finally, in Section IV we draw broad conclusions and suggest remaining questions and next steps.

[8]Fish catches are inherently divisible; but other ecosystem stocks, such as the stratosphere, are not; this can create further problems.

chapter two

Dynamic systems modeling

Robert Costanza and Matthias Ruth

Contents

How can we use the framework of Chapter 1 to build models that help us understand the complex relationships in ecological and human systems? There are many approaches but here we use conceptual and quantitative models to address the problems of scale and uncertainty in complex systems. In Sections II and III, we construct multi-scale models of ecosystems and the human systems that depend on them to inform us about regimes of sustainable resource management.

In environmental systems, nonlinearities and spatial and temporal lags are common. Some traditional analytic approaches must disregard these or treat them as anomalies. As a consequence, when interacting nonlinearities exist, traditional approaches are limited in providing insights for important management decisions. We need new modeling approaches to identify, collect, and relate the information that is relevant for understanding those systems, to make consensus building an integral part of the modeling process, and to guide management decisions (e.g., Costanza et al., 1993).

1-56670-389-1/01/$0.00+$.50

Model building can clarify problems, highlight otherwise hidden assumptions, and make effective choices among alternative actions possible. We build mental models daily for virtually all our decisions by abstracting from observations and relating relevant facts to each other. Language itself is a form of mental modeling (Pinker, 1995). For many everyday decisions, informal—even subconscious—mental models are sufficiently detailed and accurate to be reliable. Our experiences with these models are passed on to others through verbal and written accounts that frequently generate a common group understanding of the workings of any system.

To build mental models, we simplify our representations of systems in highly specific ways. We base most of our mental modeling on qualitative rather than quantitative relationships: we linearize the relationships among system components; we disregard temporal and spatial lags; we treat systems as isolated from their surroundings; and we may limit our investigations to the system at equilibrium. When problems are complex, however, and especially when quantitative relationships, nonlinearities, and time and space lags are important, our mental models may need to be supplemented.

Statistical approaches based on historical or cross-sectional data are often used to quantify system relationships. Advances in statistical methods have significantly improved our ability to deal with multiple, co-varying relationships, to test alternative models, and to test for causality (Granger, 1969, 1993). Nonetheless, a statistical modeling exercise can only provide insight into the empirical relationships over a system's history or at a point in time; they are by definition data-driven. They are of limited use to explore a system's future development under alternative management schemes, for example, because such alternatives may include decisions that have not been chosen in the past so their effects are not captured by existing data.

2.1 *Purposes of models*

Models, like maps, have many possible purposes and uses. No one map or model is right for the entire range of uses (Levins, 1966; Robinson, 1991). Models and maps are crude—but in many cases absolutely essential—abstract representations of complex territory, whose usefulness can best be judged by their ability to help solve the problems faced. When we model ecological and social systems, our purposes can range from developing simple conceptual models in order to provide a general understanding of system behavior, to detailed realistic applications aimed at evaluating specific policy proposals. It is inappropriate to judge this whole range of models by the same criteria.

Models can be tools used for building consensus in contentious and highly uncertain environmental problems (Section III). Interested parties can identify what they perceive to be the key variables, the direction and strength of relationships among variables, and the important areas of uncertainty. A model can take the best available information and generate a range of possible outcomes that are bound by uncertainty about the ecological and institutional constraints identified by the actors.[1] However, we repeat our caution of the Preface: models are essential for policy evaluation, but they can be misused. Robinson (1992) noted that there is " . . . the tendency to use such models as a means of legitimizing rather than informing policy decisions. By cloaking a policy decision in the ostensibly neutral aura of scientific forecasting, policy-makers can deflect attention from the normative nature of that decision."

2.2 Classifying different types of models

Three criteria are used to classify and evaluate models (Holling, 1964; Levins, 1966; Costanza et al., 1993). Realism describes the degree to which model representation of system behavior reflects observed behavior. Precision describes the degree to which a model represents behavior in a quantitative, exact, and repeatable way. Generality describes the degree to which a single model can represent a broad range of systems. No single model can maximize all three of these goals, and the choice of which objectives to pursue depends on the fundamental purposes of the model.

2.2.1 High-generality conceptual models

Conceptual models describe (usually qualitative) relationships between a few important variables. They simplify relationships and/or reduce resolution, thereby gaining generality at the expense of realism and/or precision. For example, the "ecological economy" model of Brown and Roughgarden (1992) contains only three state variables (labor, capital, and "natural resources"), and the relationships between these variables are highly idealized. But the purpose of the model was not high realism nor precision but rather to address some basic, general questions about the limits of economic systems in the context of their dependence on an ecological life-support base.

Simple linear and non-linear economic and ecological models tend to have high generality but low realism and low precision (Clark and Munro, 1975; Brown and Swierzbinski, 1985; Lines, 1989, 1990; Kaitala and Pohjola, 1988). Other high-generality models include Holling's "4-box" model (Holling, 1987), Folke et al.'s (1994) three scenarios of the future, most

[1]For example, the conceptual model of Folke et al. (1994) attempts to build consensus between ecologists and economists by describing the "ecological Plimsoll line." That line identifies the range of uncertainty about the quantity and quality of natural and human capital that is possible and desirable as envisioned by the two groups. Similarly, the dynamic simulation model of Hall and Hall (1993) generates a range of ways to manage the water resources of Flathead Lake, Montana, based on the needs of utilities, landowners, sport fishers, and Native Americans. That model includes the effects of uncertainty about the hydrology and ecology of the region and the conflicting demands of the users to build a consensus about how to manage the water flow through the region.

conceptual macroeconomic models (Keynes, 1936; Lucas, 1975), economic growth models (Solow, 1956), and "evolutionary games" approaches (Dugatkin and Reeve, 1998).

2.2.2 *High-precision analytical models*

High precision (quantitative correspondence between data and model) may require the sacrifice of realism and generality. One strategy here is to keep resolution high but to simplify relationships and deal with short time frames.[2] Ecologists sometimes identify one or a few properties that characterize the system as a whole (Wulff and Ulanowicz, 1989). For example, Hannon and Joiris (1987) used an economic input-output model to examine relationships between biotic and abiotic stocks in a marine ecosystem and found that this method allowed them to show the direct and indirect connection of any species to any other and to the external environment in this system at high precision (but low generality and realism). Also using input-output techniques, Duchin's (1992) aim was to direct development of industrial production systems to efficiently reduce and recycle waste in the manner of ecological systems. Large econometric models (Klein, 1971) used for predicting short-run behavior of the economy belong to this class of models; they are constructed to fit existing data as closely as possible (at the sacrifice of generality and realism).

2.2.3 *High-realism impact analysis models*

High-realism models seek to represent accurately the underlying processes in a specific system rather than precisely matching quantitative behavior or being generally applicable. Dynamic, non-linear, evolutionary systems models at moderate to high resolution generally fall into this category. Coastal physical-biological-chemical models (Wroblewski and Hofmann, 1989) used to investigate nutrient fluxes contain large amounts of site-specific data but fall into this category, as do micro models of behavior of particular business activities. Another example is Costanza et al.'s (1990) model of coastal landscape dynamics that includes high spatial and temporal resolution and complex, non-linear process dynamics. This model divided a coastal landscape into 1 km^2 cells, each of which contained a process-based, dynamic ecological simulation model. Flows of water, sediments, nutrients, and biomass from cell to cell across the landscape were linked with internal ecosystem dynamics to simulate long-term successional processes and responses to various human impacts in a very realistic way. But the model is very site-specific (low generality), and of only moderate numerical precision.

[2]Some models strive to strike a balance between mechanistic small-scale models that trace small fluctuations in a system and more general whole-system approaches that remove some of the noise from the signal, but do not allow the modeler to trace the source of system changes.

2.2.4 Moderate-generality and moderate-precision indicator models

In many systems models, the desired outcome is accurate determination of the magnitude and direction of change; these models trade off realism for some moderate amount of generality and precision. Some econometric models fall into this category; for example, Cleveland (1991) developed a model that describes the factors determining the long-run cost of oil production. The results are reasonably precise and the specification of the model describes a general relationship between resource depletion and technical change that can be applied to other resources. Other examples include aggregate measures of system performance such as standard GNP, environmentally adjusted net national product (or "green NNP") that includes environmental costs (Mäler, 1991), and indicators of ecosystem health (Costanza et al., 1992). The microcosm systems employed by Taub (1989) allow some standardization for testing ecosystem responses and developing ecosystem performance indices. Taub (1987) notes, however, that many existing indicators of change in ecosystems are based on implicit ecological assumptions that have not been critically tested for generality, realism, or precision.

2.3 Resolution and predictability

Stocks and flows can vary greatly across space and over time and therefore it may be difficult for users to perceive what predictability exists. Many, if not most, systems contain significant non-linear relationships. Non-linearity raises interesting questions about the influence of resolution (including spatial, temporal, and component) on the description of data and on the performance of models, particularly predictability. The relationship between the number of components included and the predictability of models is an important input to model design.[3] Considerable effort has been directed toward more formal measures of predictability. The predictability of resources, not just their variability or heterogenity, is very important to all resource users because a predictable resource is generally easier to utilize.

One can define predictability for a whole host of resources. One can also distinguish between "data," or descriptive predictability within a data series, and "model" predictability, the comparison of a model to existing data. For example, collected rainfall data will exhibit a certain descriptive predictability. Particular models of rainfall will also exhibit differing abilities to predict rainfall data (model predictability). In the spatial domain, one can define two

[3]Hofmann (1991) discusses this concern in the context of scaling coastal models to the global scale. It is difficult to use aggregate models that integrate over many details of finer resolution models because the aggregated models may not be able to represent biological processes on the space and time scales necessary. Hofmann suggested that detailed models that were "coupled," in which the output of one model becomes the input for another, may be a more practical method for scaling models to larger systems. The Patuxent Landscape Model, discussed in Chapter 9, is such a model.

types of predictability, (1) spatial auto-predictability (P_a) which is the reduction in uncertainty surrounding the state of a pixel in a scene, given knowledge of the state of adjacent pixels in that scene; and (2) spatial cross-predictability (P_c) which is the reduction in uncertainty about the state of a pixel in a scene, given knowledge of the state of corresponding pixels in other scenes. P_a is a measure of the internal pattern in the data while P_c is a measure of the ability of a model to represent that pattern. For example, cross-predictability (P_c) can be used for pattern matching and testing the fit between map scenes. While P_a generally increases with increasing resolution (because more information is being included) P_c generally falls or remains stable (because it is easier to model aggregate results than fine grain ones). Costanza and Maxwell (1994) analyzed the relationship between spatial resolution and predictability and found that while increasing resolution provides more descriptive information about the patterns in data, it also increases the difficulty of modeling those patterns accurately. Thus, we can define an optimal resolution for a particular modeling problem by balancing the benefit in terms of increasing data predictability (P_a) as one increases resolution, with the cost of decreasing model predictability (P_c). Figure 2.1 shows this relationship in generalized form. There may be limits to the predictability of natural phenomena at particular resolutions, and "fractal-like"

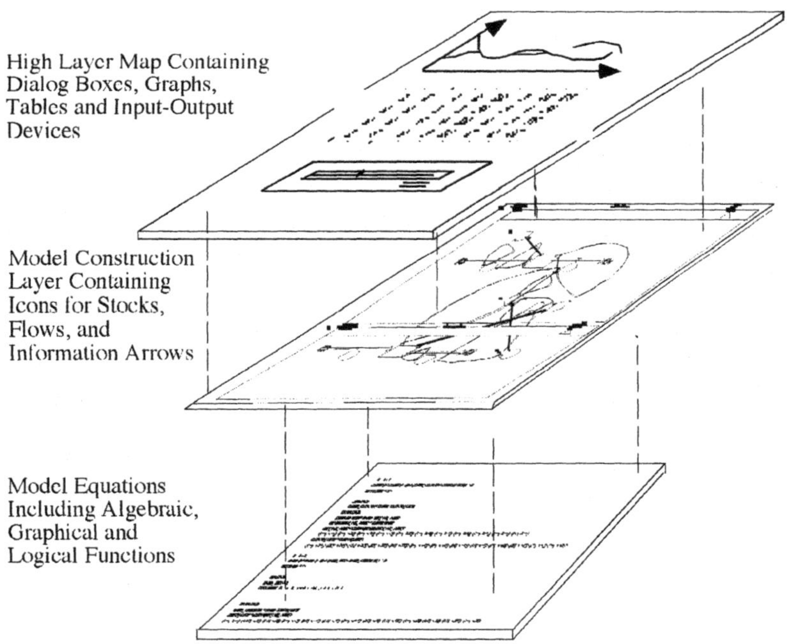

Figure 2.1 STELLA II Modeling Environment.

rules that determine how both "data" and "model" predictabilities change with resolution.

2.4 Models for consensus building

Various forms of computer models for scoping and consensus building have been developed for business management applications (Chapter 8; also Roberts, 1978; Lyneis, 1980; Morecroft et al., 1991; Vennix and Gubbels, 1994; Morecroft and van der Heijden, 1994; Westenholme, 1990, 1994; Senge and Sterman, 1994). Previously, emphasis was placed on the provision of computer hardware and software to support group communication (Kraemer and King, 1988). Recent trends are to facilitate problem-structuring methods and group decision support (Checkland, 1989; Rosenhead, 1989; Phillips, 1990). The use of computers to structure problems and provide support for group decisions has been spurred by the recognition that in complex decision settings the bounds on human rationality can create persistent judgmental biases and systematic errors (Simon, 1956, 1979; Kahnemann and Tversky, 1974; Kahnemann et al., 1982; Hogarth, 1987). Dynamic systems modeling is increasingly being used as a tool for (1) closing spatial and temporal gaps among decisions, actions, and results; (2) assessment of relationships among decisions, actions, and results; and (3) the facilitation of learning that requires that cause and effect are related in space and time.

Dynamic systems modeling is increasingly being used to help avoid judgmental biases and systematic errors in business management decision-making (Senge, 1990; Morecroft, 1994). It has also penetrated (to a lesser extent) the discussion of environmental investments and problems. Both areas of application of dynamic systems modeling have significantly benefited from the use of graphical programming languages.[4] With their relative ease of use, these graphical programming languages offer powerful tools for intellectual inquiry into the workings of complex ecological-economic systems (e.g., Hannon and Ruth, 1994). There are various graphical programming languages available that are specifically designed to facilitate modeling of nonlinear, dynamic systems. Among the most versatile of these languages is the graphical programming language STELLA II™ (Costanza, 1987; Richmond and Peterson, 1994; Hannon and Ruth, 1994), which runs in both Macintosh and Windows environments. The models of Section II are STELLA models.

---

[4]One of the main strengths of these programming languages is enabling scientists and decision makers to focus and clarify the mental model they have of a particular phenomenon, to augment this model, elaborate on it, and then to do something they cannot otherwise do: to run the model and let it yield the inevitable dynamic consequences hidden in their assumptions and their understanding of a system.

A STELLA II dynamic systems model consists of three communicating "layers" that contain progressively more detailed information on the structure and functioning of the model (Figure 2.1). The high-level mapping and input-output layer facilitates user interaction through input and output devices. This layer is most appropriate for defining the structure of the model and enabling non-modelers to do important tasks such as grasp the model structure, run the model interactively, and interpret results. The ease of use of the model at this aggregate level of detail thus enables individuals to become intellectually (and emotionally) involved with the model (Peterson, 1994).

Models are constructed in the next lower layer. Here, the symbols for stocks, flows, and parameters are chosen and connected with each other. STELLA II represents stocks, flows and parameters, respectively, with the following three symbols:

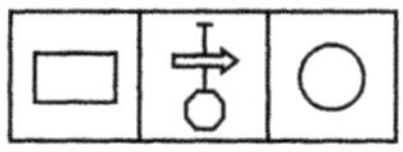

Icons can be selected and placed on the computer screen to define the main building blocks of the computer model. The structure of the model is established by connecting these symbols through "information arrows."

Once the structure of the model is laid out on the screen, initial conditions, parameter values, and functional relationships can be specified by simply clicking on the respective icons. Dialog boxes appear that ask for the input of data or the specification of graphically or mathematically defined functions.

It is also easy to generate model output in tabular or graphical form by choosing icons. With the use of "sliders" a user can also respond to the model output by choosing alternative parameter values as the model runs. Subsequent runs under alternative parameter settings and with different responses to model output can be plotted in the same graph or table.

Thus, the modeling approach is dynamic, not only with respect to the behavior of the system itself but also with respect to the learning process that is initiated among decision makers as they observe the unfolding of the system's dynamics. Subsequent runs under alternative assumptions help determine aspects of the system that may be insufficiently understood, may prompt further model development, and can help stakeholders to systematically evaluate their perceptions of the workings of a system. The process of modeling experiments on the computer gives model users the opportunity to investigate the implications of their assumptions for the system's dynamics and to assess their ability to make the "right" decision under alternative assumptions.

The lowest layer of the STELLA II modeling environment contains a listing of the graphically or algebraically defined relationships among the system components together with initial conditions and parameter values. These equations are solved in STELLA II with numerical techniques. The equations, initial conditions, and parameter values can also be exported and compiled to conduct sophisticated statistical analyses and parameter tests (Oster, 1996) as well as to run the model on various computing platforms (Costanza et al., 1990; Costanza and Maxwell, 1991).

section II

Models

chapter three

Human-ecosystems interactions: a basic dynamic integrated model

Bobbi S. Low, Elinor Ostrom, Robert Costanza, and James A. Wilson

Contents

Managing our use of important ecosystem resources can clearly be problematic (e.g., McCay and Acheson, 1987; Ludwig et al., 1993; Jansson et al., 1994). We see conflict today over the fact that many resources are "mined"—treated as though they were a non-renewable resource in which the only "rational" strat-

1-56670-389-1/00/$0.00+$.50

egy is to "get mine and get out." These include numerous oceanic and coastal systems including cod, whales, and lobsters. Managing such resource systems can be extremely difficult for a number of reasons including (1) both resource stock and harvesters may cross boundaries; (2) conducting an accurate census of some resource stocks is difficult-to-impossible; (3) varied actors may be in conflict over resource use and may not be able to agree on rules; and (4) the stock may, like whales harvested in the open ocean, constitute a common-pool resource with all the attendant problems (e.g., Ostrom, 1990; Ostrom et al., 1999). The problems are particularly likely when several jurisdictions are involved or when a single rule system is applied to ecologically-diverse areas.

How can the framework of Chapter 1 help design effective resource management systems? Well-developed and accepted models exist for resource-use modeling (Chapter 6; also Conrad and Clark, 1987; Hilborn and Walters, 1992) and some are explicitly tailored for fisheries stocks, which we explore in this chapter. These models can be dynamic yet we see several advantages to the integrated approach we propose here. First, none of the existing models explicitly integrates (possibly diverse) human decisions in a clear feedback system linking both ecosystems and human decision systems. Second, even the dynamic models are not tailored to analyze the impacts of unpredictability—both through ecological events that may influence stock populations independent of human action and through error in our censuses of stock (e.g., fish stocks are notoriously difficult to assess accurately). Using an integrated model, we can explore these effects. Interacting forces, often non-linear, can be modeled and "experiments" run to determine sensitivity to particular changes. Finally, the system we develop here allows explicit linking and exploration of physical and human systems that may differ and have transfer, of either harvesters or stocks (this chapter, Chapter 4, and Chapter 9). Thus, one can test some important broad issues in a linked model including spatial diversity, stock and harvester transfers, and rule robustness in the face of uncertainties.

To be operational, a model based on the framework of Chapter 1 must not only define stocks, flows, attributes, and interactions, but should also specify (Figure 3.1):

1. The ecological carrying capacity of the system
2. The degree to which external influences are predictable
3. The regeneration rate (population growth and mortality rates) of the resource
4. The transfer rate of resource users (or harvesters) from one spatial unit to another
5. The consumption rate(s) of resource users

This model links human decision systems and ecosystems in parallel ways (Chapter 1; Cleveland et al., 1996). We construct the models of this and subsequent chapters using STELLA; standard notation is presented in each chapter, and the STELLA equations are appended. Appendix 3.1 gives the STELLA equations relevant to Figures 3.1 and 3.4. It includes options (e.g.,

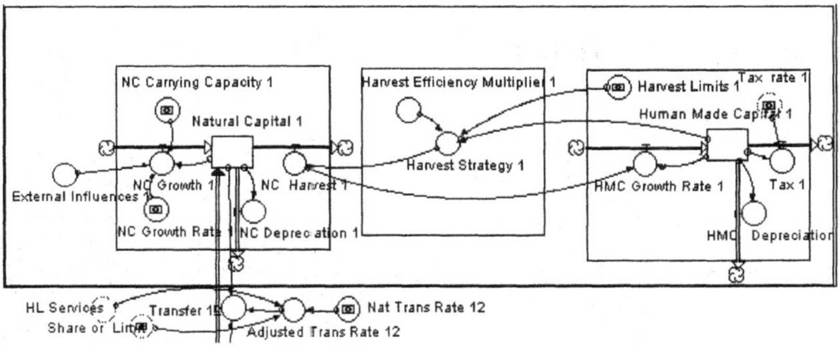

Figure 3.1 A simplified model of an ecosystem in which some resource stock is utilized by humans. Stocks, flows, and controls are further defined in the text. This is the one-unit model.

shifts in tax rates) not explored here.

Initially, the model is extremely simple. First we establish its basic behavior as an isolated system with no uncertainties and compare four harvest rules: percent carrying capacity, percent population (or standing crop biomass), open access, and profit maximizing sole owner. Second, we test the rules in the face of ecological uncertainties that affect the stock. Finally, we model three linearly-linked systems, as in a series of coastal fisheries, and test the rules in the face of ecological uncertainties and various stock transfer rates. Chapter 4 tests three fully- (rather than linearly-) linked systems and explores the effects of panmictic versus meta-population structure.

3.1 *The model*

3.1.1 *Ecosystem sector*

The ecosystem sector (Figure 3.1; Appendix 3.1) contains one state variable, labeled natural capital, that represents the biomass of a fish stock. Natural capital (NC) grows over time at some intrinsic rate, is limited by the carrying capacity of the ecosystem *(K)*, and may be influenced by external influences (temperature shifts, pollution) that can be unpredictable in their timing and extent:

$$NC_{growth} = NC * ExtInfl * NC_{growth\ rate} * [1 - NC/K] \tag{3.1}$$

Equation (3.1) simply represents the stock's logistic growth, as it is influenced by carrying capacity and external events.

Stocks are reduced not only by deaths and emigrations, but also by human harvest (Figure 3.1). The overall equation for natural capital, where $NC_{depreciation} = (NC_{depreciation\ rate}) * NC$, is:

$$D(NC)/dt = NC_{growth} - NC_{depreciation} - \text{Transfer} - NC_{harvest} \quad (3.2)$$

We set the initial value of natural capital at the start of the simulation to equal the assigned long-term carrying capacity. The transfer value is zero in the first simulations; it is active only when stock can move from one system to another.

The logistic growth of dynamic Equation (3.2), with NC harvest and Transfer set to zero, produced the standard parabolic recruitment curve of most bioeconomic models (Chapter 6; Hilborn and Walters, 1992). Under these assumptions the actual carrying capacity is $x* = K(1 - (NC_{depreciation\ rate} / (ExtInfl * NC_{growth\ rate})))$, and the maximum sustained yield (MSY)—the maximum value of the growth curve—is:

$$x_{MSY} = ((K * NC_{growth\ rate} * ExtInfl)/4) * (1 - (NC_{depreciation\ rate} / (ExtInfl * NC_{growth\ rate})))^2 \quad (3.3)$$

where *K* is the cHMCdepreciation − Tax.

3.1.2 *Human-ecosystem interaction sector*

The interaction sector has no state variables, only two-way flows between ecosystems and human systems (Figure 3.1, Appendix 3.1). In this initial application we focus on two relationships, harvest efficiency and harvest strategy. In the models of this chapter, we assume that the amount harvested is proportional to both human-made capital (HMC) and the size of the resource population (NC). We call this proportionality factor total efficiency (TE) and set it equal to 0.007 in these simulations.

In the absence of any externally-imposed limits, the amount harvested is HMC*NC*TE. We call the multiplier NC*TE the "harvest efficiency" (HE). In modeling resources for which there is no search problem (i.e., when harvest success is unrelated to resource abundance), we would use a constant HE, independent of the value of NC.

We now consider four harvest rules: maximum sustained yield (MSY), percent population or biomass, open access, and profit maximizing sole owner. Below, we will explore some impacts of over-harvesting, but at first all fishers will obey the harvest rules completely. The MSY rule, one of four rules we explore in this chapter, sets the harvest limit, HL, at a constant percentage of the carrying capacity, K (Chapter 6). The harvest (HS) is then given by:

$$HS_1 = \min\{HMC*HE, HL\} \quad (3.4)$$

Thus if the potential harvest, HMC*HE, is less than the harvest limit, the potential harvest is taken; otherwise the harvest limit HL is taken.

The "percent population" rule simply sets the harvest limit at a specified percentage of the current population or standing crop biomass (Appendix 3.1). This rule requires highly accurate censusing of the stock at appropriate times to predict sustainable harvest; this may include staggered collection efforts due to seasonality of stock reproduction, for example. With the values used in the simulations here, harvesting 28.5% of the existing stock was found to maximize HMC.

In the open access regime, each actor responds to his/her current profit, and keeps harvesting so long as the profit is not negative. Fishers will continue to enter in this system until total profits are zero; they are short-term profit maximizers insensitive to trends, with no incentive for any restraint.

Under the "profit maximizing sole owner" rule, the sole owner controls access to the natural capital and harvests it at a rate that maximizes long-term profit. The sole owner's time horizon is longer than that of the open-access harvesters. In the simulations of this chapter, the sole owner's decisions are based on the 5-year trend (slope) of profits, rather than on the most recent profits. We do this because delays from feedback in biological systems can cause the most recent profits to give a false signal. A 5-year trend gives a slower response, but tends to find the sustainable profit-maximizing harvest level more reliably.

3.2 *Rules and sustainability in a single-ecosystem model*

For convenience in these STELLA runs, K is set to 10,000 and the initial value of natural capital is set to the carrying capacity. We allow the system to run for a maximum of 200 periods (years), and, unless otherwise specified, do 100 runs. When a system does not sustain itself (defined here as having natural capital in excess of 20 units after 200 rounds), we track the number of years until collapse of the system. This is the equivalent of gathering empirical data on 100 fisheries for up to 200 years each. For this system, we explore:

1. how harvest limits influence the stocks of natural and human-made capital
2. how the growth rate of natural capital influences the sustainability of various harvest limits
3. how (1) and (2) interact
4. how stochastic fluctuations of ecological influences (Figure 3.1) independent of human harvest affect sustainability. In these runs, fluctuations are randomly generated around a mean of 1 (no fluctuations), so we are exploring the effects of changes in the range of unpredictable variation in external factors

In the first three analyses, we focus on deterministic relationships to explore the underlying curvilinear structure affecting long-term survival of natural capital. Then we model stochastic environmental fluctuations that affect the growth of natural capital, mimicking some complexities of empirical data.

3.2.1 *Harvest rules and stock growth rate in a single-system deterministic model*

Local, regional, or national authorities frequently impose an upper bound on the quantity of harvest that may be taken during a defined time period. One such calculation that is frequently used is that of maximum sustained yield (MSY). We see easily from Equation (3) and Chapter 6 that if one assumes a constant harvest rate, MSY equals 0.2401 K under our assumptions here of a stable population ($NC_{growth\ rate} = 0$, death rate = 0.2, and no external fluctuations (ExtInfl = 1)). For a series of growth rates ($-0.1 - +0.1$), we varied the authorized harvest rate from 15% to 30% of K. One can see in Equation (3) that MSY is an increasing function of growth rate: $x_{MSY}/NC_{growth\ rate} = (ExtInfl*K)/4 * (1 - (NC_{depreciation\ rate}/ (ExtInfl * NC_{growth\ rate})))^2 > 0$. This is no surprise; rapidly growing stocks can sustain heavier exploitation. In all systems, exceeding MSY caused the collapse of natural capital and the system, and the greater the excess, the more rapid the collapse. As we note below, however, in actual practice MSY is problematic.

3.2.2 *Growth and harvest rates with ecological perturbations*

The intrinsic growth rate of an exploited stock clearly affects both sustainability and the relative effectiveness of different management strategies. Extrinsic fluctuations are mimicked here by stochastic shifts affecting stock levels by ± 10% or ± 50%. In a series of runs tracking both natural and human-made capital in a single-unit system, the strongest predictor of the amount of natural capital at the end of a run was harvest limits (d.f. = 2,98, $r^2 = 0.63$, $p < 0.0001$), although the growth rate of natural capital was also highly significant. Human-made capital was most affected by harvest limits (d.f. = 2,98, $r^2 = 0.27$, $p < 0.00001$) while the growth rate of natural capital had less of an impact on HMC ($p < 0.032$). The strongest predictor of years until collapse of the system was harvest limits (d.f. = 2,98, $r^2 = 0.49$, $p < 0.00001$) although the natural capital growth rate had some influence ($p < 0.009$).

This exercise highlights some of the widely-recognized difficulties in using the concepts of K, carrying capacity, and MSY (Conrad and Clark, 1987; Clark, 1990). In real-world situations, the use of carrying capacity ignores information. Figure 3.2 highlights an additional management problem. Up to the sustainable harvest limit, human-made capital increases linearly with the harvest limit, and the slope of this increase is steeper than the decline of natural capital. That is, up to the sustainable limit, there are relatively great increases in human-made capital for small decreases in natural capital–an obvious incentive to continue taking more. Further, resource users are able to measure changes in human-made capital easily, and more reliably than changes in natural capital. When sustainable limits are exceeded even slightly in this model, there is a dramatic collapse of both kinds of capital. Thus, there is little "early warning" (Gulland, 1977) that the limits of sustainability are being approached, nor is there feedback to allow regulators to react to changes in stock size.

These results (Figure 3.2) look similar to difficulties encountered in the field. High harvest limits (even when no cheaters exist) can build human capital at the expense of natural capital—and the pattern of the two capitals gives little prior warning of impending collapse. Indeed, many resource-use systems fail relatively abruptly. These models do use a relatively unconstrained mechanism, in which, all human-made capital is plowed back into harvesting capacity, whereas more realistic and constrained rules (other uses of human capital at the expense of that available for harvest) should show less dramatic patterns.

3.2.3 Harvest rules and stochastic ecological perturbations

In the real world, even in isolated systems, ecological events can affect the level of natural capital (and thus the outcomes of harvest rules) independent

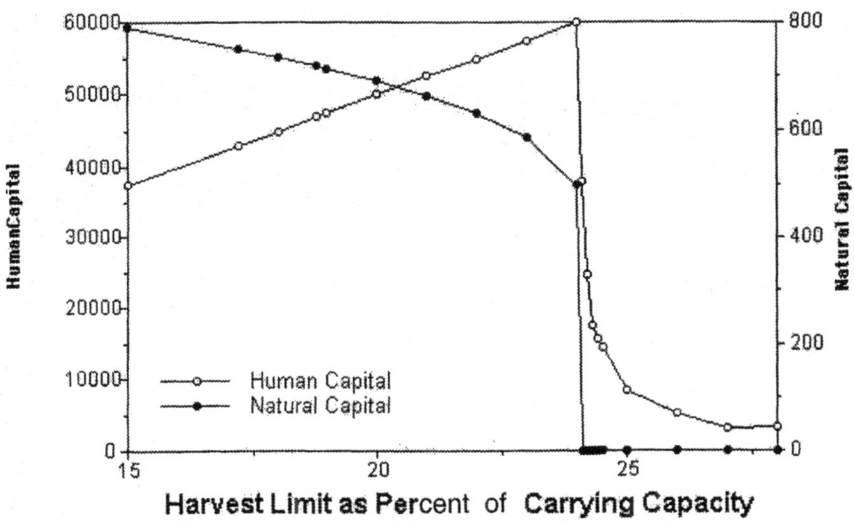

Figure 3.2 Up to the sustainable harvest limit (here, 24% of carrying capacity), profit (human capital at the end of the 200-year simulation) increases as the harvest limit increases; however, the decline in natural capital (biomass of the resource at the end of the 200-year simulation) declines less dramatically. When sustainable limits are exceeded, the collapse is rapid and dramatic. In this model, there is little "early warning" that sustainable limits are being approached.

of human activity. Consider Equation (3.3) and the discussion above: MSY is non-responsive to ecological fluctuations. The impact increases as the fluctuations increase in amplitude (Figure 3.3a). When the rule being modeled was "harvest 24% of K" and ecological impacts influenced the fish stocks by ± 10%, approximately 15% of systems maintained some natural capital by year 200, and most survived at least 100 years. With the same harvest rule but ± 50% ecological impacts, 57% of systems failed by 40 years and only 8 of 100 harvesters persisted beyond 120 years.

Other harvest rules out-performed MSY. Figure 3.3b compares the rules under perturbations of ± 50%. In contrast to the MSY rule, no sole owner system failed within 40 years, 57% of sole owner systems persisted past 120 years, and nine were functioning at 200 years. Even open-access rules out-performed MSY when harvesters were sensitive to current profit. All systems using the rule "50% of stock" were functioning capably at 200 years. Note, however, that this rule still retains assumptions that are unrealistic in the real world, including that we have captured the dynamics accurately, that stock populations can be sampled with accuracy, and that no delays (lags) exist for any reason (see Chapter 4).

3.3 *Spatial heterogeneity*

These results are consistent with our understanding of the expected behavior of an isolated socio-economic system (Chapter 6). However, this approach explicitly assumes no significant interactions with other populations, and stock populations are rarely isolated in this way.

Consider for example, a series of coastal fisheries in which fish can move from one area to another (Figure 3.4) but where all harvesters are local harvesters, even though harvesters often can also move from one area to another (see Chapter 4). The Maine lobster fisheries have a complex set of conditions that have caused difficult management conflicts for some years, for reasons both biological and political. Only older mature lobsters are marketable in Maine, and non-marketable lobsters are marked with a "V" notch in the tail. Non-marketable lobsters include immature lobsters, gravid females, and very large lobsters who are presumed to provide a reproductive stock as a cushion. In Maine, no one can legally market these marked individuals. However, because no such restrictions apply in New Hampshire and Massachusetts, lobstermen from these states could line up at the state boundaries. The inability to enforce the v-notch rule across state boundaries, reduces the conservation incentive of Maine lobstermen. Outsiders are free riders, whose actions mean a more compressed size/age distribution in all areas, and thus a greater risk of recruitment failure when ecological perturbations occur. This represents a mismatch between rules and ecological realities because there is biological transfer across systems, local harvest rules, and no higher-order control (see Chapter 10). This problem is already being addressed and these circumstances will probably end before 2001.

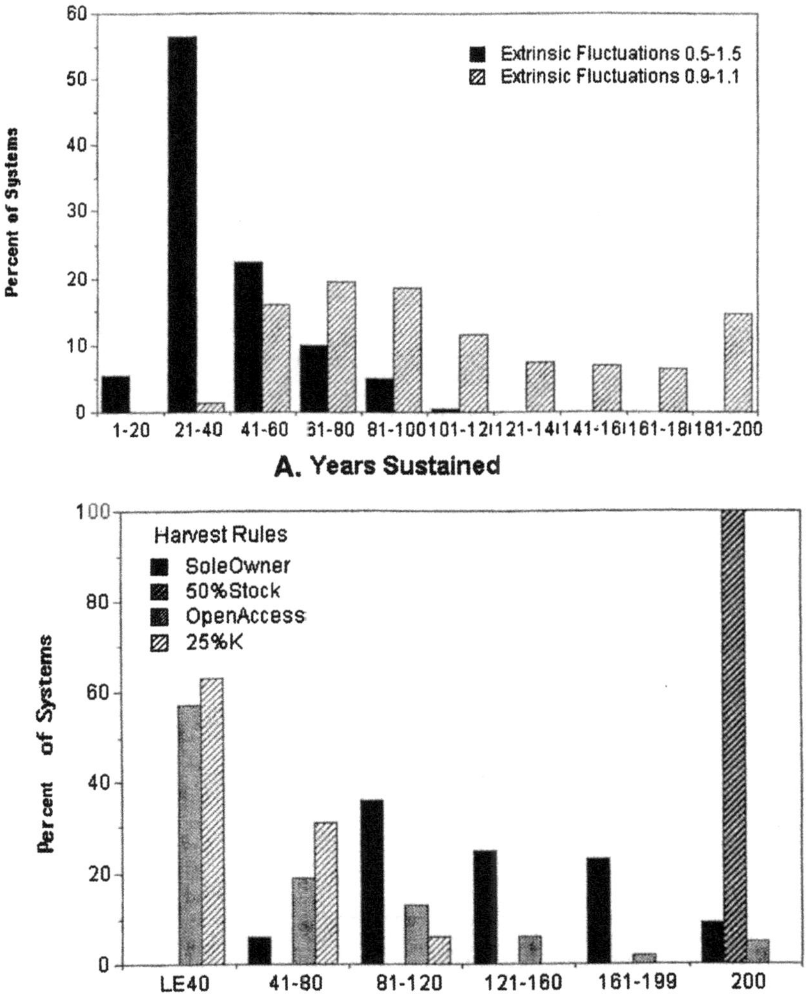

Figure 3.3 (a) The sustainable harvest limit in any system will be influenced not only by the growth rate of the resource stock, but by fluctuations in the extrinsic ecological factors that influence the stock level (e.g., any fluctuation causing deaths or heightened recruitment of the stock). With a harvest limit of 24% of carrying capacity (sustainable if there are no extrinsic fluctuations), some systems fail. Systems fail more frequently when extrinsic fluctuations are of greater magnitude (here, extrinsic factors affect the population causing fluctuations from 50% to 150%, average = 100%) than when fluctuations are more limited (90%–110%, average = 100%). (b) The four harvest rules tested are differentially vulnerable to extrinsic fluctuations in stock populations of ± 50%: "percent stock" and "sole owner" systems proved more robust than "% carrying capacity" and open access. Note, however, that complications such as miscounts of stock populations and lags in effect of fluctuations are not included, so all systems here perform better than is likely in real world systems.

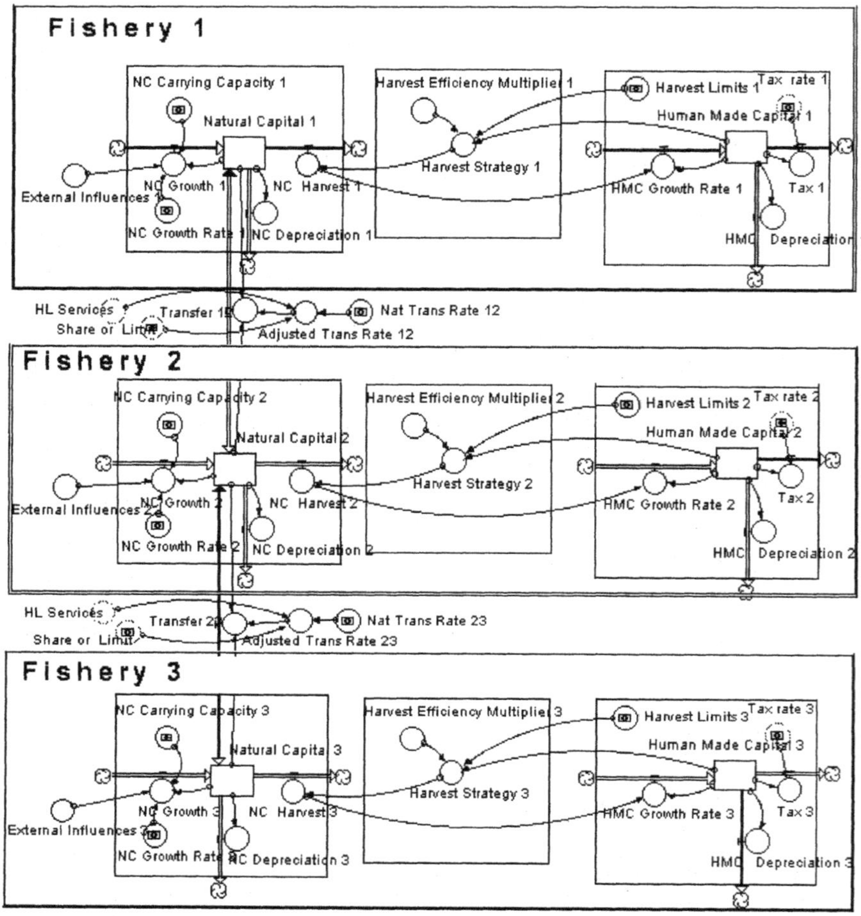

Figure 3.4 When resource systems are not isolated, both resource stocks and resource users may move among systems. Here, movement of natural capital is shown as it could occur across three coastal fisheries. Other possible movement patterns are shown in Chapter 4.

Maine fishermen took their case to the Atlantic States Marine Fisheries Commission (ASMFC), an interstate body that has authority to govern lobster conservation. The ASMFC agreed with the position of Maine lobstermen and extended the area in which v-notch and oversize protection is enforced to the entire Gulf of Maine. The designated area does not fully contain the movements of lobsters (due to political compromises within ASMFC and also because it is virtually impossible to draw a perfect ecological boundary). Nonetheless, the improvement brought about by the actions of the ASMFC substantially aligns the scale of human rules and the biologically relevant behavior.

3.3.1 *Spatial representation of multiple ecosystems*

How, and at what levels, do biological transfers among non-isolated systems affect situations such as the Maine lobster industry? Each of the spatial units (fisheries) in Figure 3.4 has an ecosystem sector, a human decision sector, and an interaction sector (see also Sanchiro and Wilen (1999) who have independently developed a remarkably parallel system). To track processes, all equations for each unit are appended with a suffix reflecting that unit (Appendix 3.1). In each spatial unit, variables such as initial natural capital, NC growth rate, and carrying capacity can differ among the ecosystem sectors, while in the human sectors, decision rules can vary.

Here, we assume the natural capital transfer rate between units (T) is proportional to the biomass differential in the two units. If the units are isolated, T = 0. A transfer rate of 50% of the differential between adjacent units will equalize the stock in the two units.

This series of experiments explores how levels of natural capital transfer between systems affect the ability of people in one unit to exceed the MSY in the absence of any ecological perturbations. Without ecological perturbations, the amount that can be taken sustainably by any unit depends on the carrying capacities in the units, the natural capital growth rates in the units, and the transfer rate. If the transfer rate equalized all differences (e.g., a fully panmictic population, with spatial equilibration), then for a three-unit set of ecosystems, with the settings used above, the MSY would be 24% of the total carrying capacity (i.e., 0.24*3000, or 720). The units would be totally interdependent with regard to harvest limits. Most systems, while not completely isolated, have some natural transfer across sub-systems. When such incomplete transfer occurs, what any unit can take sustainably is a complex function of its own harvest, the harvest of other units, and the transfer rate.

3.3.2 *Impact of cheating*

What will happen in our 3-system coastal fisheries, when fishers in the central system take more than a sustainable amount–when they cheat? How is this affected by the transfer rate of stocks across systems, and by the takes of fishers in the other two systems?

When the transfer rate is very low, such a unit will be unsustainable when it exceeds 24% even slightly (Figure 3.5). When transfer rates are higher and the harvests in neighboring "conserving" units are low, highly exploitative harvesters, who take more than would be sustainable in an isolated system, receive some protection from this combination of transfer rates and conserving behavior by neighbors.

How much conservation is needed in other units in order to sustain overexploiters? For even very low rates of natural capital transfer, a complex interaction occurs among *T* and the harvest rates of the conserver and the exploiter systems. To reflect this complexity, we construct a "Harvest Pressure Bias" (HPB) index:

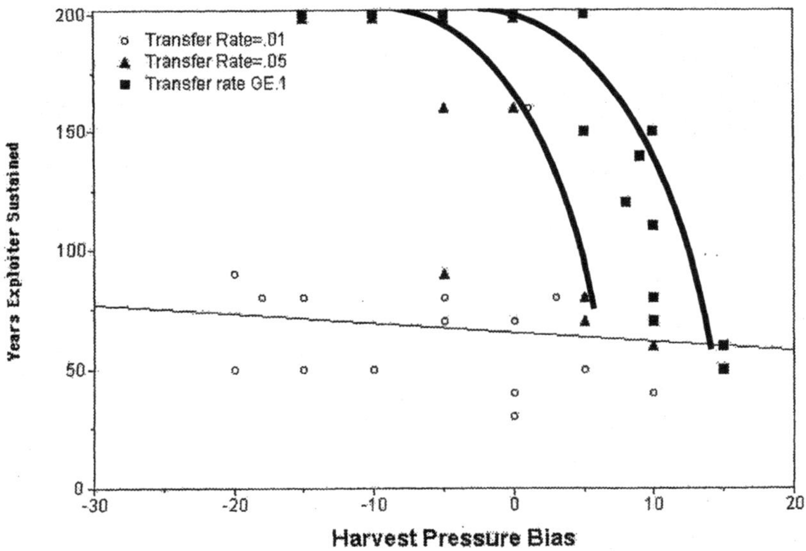

Figure 3.5 In a system such as that shown in Figure 3.4, the transfer rates of natural capital and the harvesting rates of all units interact. Here, combined harvest rates are reflected by the "harvest pressure bias" ([exploiter harvest rate-MSY_e] − [MSY_c-conserver harvest rate]); this number increases as harvest rates increase in any system. When natural capital transfer rates are low, the failure of the highest-harvesting unit is relatively independent of the total harvesting bias, and "exploiters" fail at approximately the same harvest levels as for isolated systems. At higher transfer rates, units that harvest heavily can be protected and persist throughout the run (200 years), up to some harvest bias (i.e., when other units harvest near the MSY, or the exploiter takes ≥ 26% of the carrying capacity). This (inadvertent) support of an exploiting unit is shown by the flat portion of the high-transfer rate curves. The higher the natural transfer rate, the larger the region over which such protection exists. When the natural capital finally collapses, however, the "conserving" units collapse along with the exploiting unit (Figure 3.6).

$$\text{HPB} = ([\text{exploiter rate} - \text{MSY}_e] - [\text{MSY}_c - \text{conserver rate}]) \quad (5)$$

The HPB index represents the ranges of conservationist-versus-exploiter harvests that will leave the entire linked system's natural capital sustainable. In our runs, this measure ranged from −20 to +15, depending on the natural capital transfer rate (Figure 3.5). When the transfer rate was 5%, a HPB greater than −5 meant collapse. When the transfer rate was 10%, a HPB of +5 could be reached before collapse. When the harvest limits are exceeded, the time over which an exploiting unit can be sustained drops precipitously and the other units may face ruin as well.

Theoretically, the combination of conserving neighbors and high transfer rates could shield "cheating" (over-harvesting) harvesters. In most cases, however, neighbors in conserving units will learn of the situation, as in the Maine lobster example above. If conserving units then respond by raising their harvest rather than shielding the free riders, the entire system will collapse. A real problem exists in matching local and supra-local rules to ecological realities of transfer rates, as well as to more obvious phenomena like carrying capacity and growth rate.

In reality, the combination of high transfer rates and conservative neighbors affords protection for exploiters only if there are no ecological perturbations, a seldom-met condition. To explore the effects of perturbations as they interact with harvest limits and transfer rates, consider the following conditions:

1. Transfer rate = 0.1, range of perturbations = 0.9–1.1(NC). (On average, stochastic variations will have no effect on natural capital level, but in any one random year the variations could affect the natural capital amount by ± 10%.) An exploiter in this system, as well as in #3 below, would persist 150 years.
2. Transfer rate = 0.1, range of perturbations = 0.5–1.5(NC). With no perturbations, an exploiter in this system and in #4 below, would persist 150 years.
3. Transfer rate = 0.2, fluctuations = 0.9–1.1 (NC).
4. Transfer rate = 0.2, fluctuations = 0.5–1.5 (NC).

Figure 3.6 shows that with conservative neighbors and a 10% natural capital transfer rate, the presence of any fluctuations reduces the probable persistence of the exploiting unit. Greater fluctuations reduce persistence more than moderate fluctuations (Figure 3.6a). When transfer rates are higher (20%; conditions # 3 and 4 above), the interactions change the response. With moderate fluctuations and a 20% transfer rate, approximately 20% of exploiters persist to 200 years. However, the 20% transfer rate interacts with high perturbations—more than 60% of exploiters fail by year 40 (Figure 3.6b).

3.4 Discussion and conclusions

The models in this chapter are very simple, compared to many more elegant formal fisheries models. Their principal advantages are that: (1) they can link more than stock and economic returns, to let us look simultaneously at human and ecological systems under multiple regimes; (2) they allow spatial modeling, with transfers across units; and (3) we can model unpredictable perturbations easily and explicitly.

It is well-known that the growth rate of natural capital interacts with harvest limits to affect sustainability (Chapter 6). But growth rate is difficult to measure accurately, as are carrying capacity and degree of perturbation in the system. Some of the variables we have examined here and found to have significant influence on sustainability, are rarely measured.

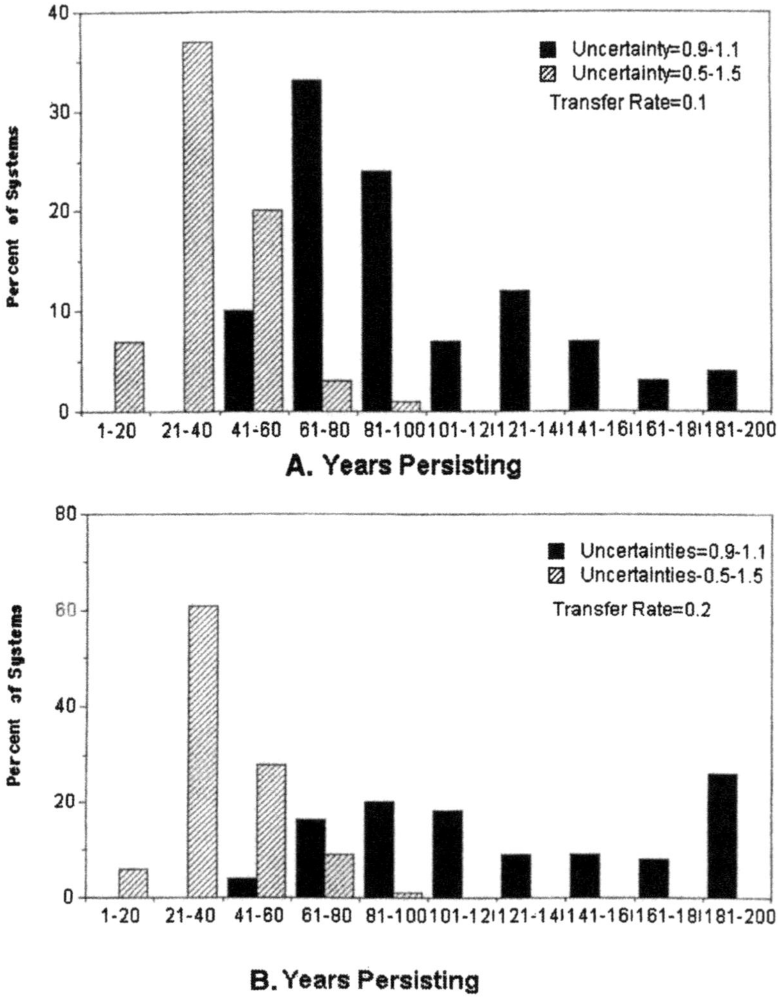

Figure 3.6 The interactions of transfer rates, total harvest pressure bias, and extrinsic ecological uncertainties mean that high transfer rates alone cannot protect exploiters. In all cases here, the total harvest pressure is 720 ($= 3 \times 240$, or the sustainable limit if all units acted identically); the exploiting unit takes 250, and the two conservative units each take 235. At both transfer rates (0.1 and 0.2), higher extrinsic fluctuations cause more failures of the exploiter. (a) When the transfer rate is 0.1 (10% of stock differential can move between units in a time period), $> 30\%$ of exploiters fail by year 40 (extrinsic fluctuations 0.5—1.5), or by year 80 (fluctuations 0.9—1.1), as opposed to persisting 150 years, the case if there were no extrinsic fluctuations. (b) The pattern is different for higher transfer rates (0.2): moderate (0.9—1.1) fluctuations and high transfer rates allow $> 20\%$ of exploiters to persist for the full 200-year run (as would be the case with no extrinsic fluctuations). However, in the face of both high fluctuations and high transfer rates, $> 60\%$ of exploiters fail by year 40—and in contrast to low transfer rate conditions, these failures have also destroyed the neighboring conservative units.

Many models of sustainability are based on long-term averages, without regard to the impact of perturbations in the system that affect natural capital levels. Yet the explorations above suggest that such perturbations can be extremely important and we suggest that it would be profitable to pay more attention to measures of variance. If sustainability is a management goal, such attention would lead managers to invoke "safe minimum standards" (Bishop, 1993), or what we call the "precautionary principle" (Low and Berlin, 1984; Costanza and Cornwell, 1992; Costanza et al., 1998; also see Chapter 10). Taking such steps would likely be unpopular, since these standards manage not for the average conditions, but avoid worst-case conditions. Since this will lower harvesters' profits in most cases, and since empirically the appropriate levels are difficult to determine, conflicts of interest are likely. As we argue in later chapters, however, the information gained by using models to do experiments may prove helpful in working through such conflicts (chapters 8, 9, and 10).

Appendix 3.1
STELLA Model Equations

Ecosystem 1:

Natural_Capital_1(t) = Natural_Capital_1(t − dt) + (NC_Growth_1 − NC__Harvest_1 − NC_Depreciation_1 − Transfer_12) * dt

INIT Natural_Capital_1 = NC_Carrying_Capacity_1

INFLOWS:
NC_Growth_1 = External_Influences_1*(NC_Growth_Rate_1*Natural_Capital_1*(1 − (Natural _Capital_1/NC_Carrying_Capacity_1)))

OUTFLOWS:
NC__Harvest_1 = harvest
NC_Depreciation_1 = Natural_Capital_1*.02
Transfer_12 (Not in a sector)
NC_Carrying_Capacity_1 = 10000
NC_Growth_Rate_1 = 1

Ecosystem 2:

Natural_Capital_2(t) = Natural_Capital_2(t − dt) + (NC_Growth_2 + Transfer_12 − NC__Harvest_2 − NC_Depreciation_2 − Transfer_23) * dt

INIT Natural_Capital_2 = NC_Carrying_Capacity_2

INFLOWS:
NC_Growth_2 = External_Influences_2*(NC_Growth_Rate_2*Natural_Capital_2*(1 − (Natural_Capital_2/NC_Carrying_Capacity_2)))
Transfer_12 (Not in a sector)

OUTFLOWS:
NC__Harvest_2 = harvest_2
NC_Depreciation_2 = Natural_Capital_2*.02
Transfer_23 (Not in a sector)
NC_Carrying_Capacity_2 = 10000
NC_Growth_Rate_2 = 1

Ecosystem 3:

Natural_Capital_3(t) = Natural_Capital_3(t − dt) + (NC_Growth_3 + Transfer_23 − NC__Harvest_3 − NC_Depreciation_3) * dt

INIT Natural_Capital_3 = NC_Carrying_Capacity_3

INFLOWS:
NC_Growth_3 = External_Influences_3*(NC_Growth_Rate_3*Natural_Capital_3*(1 − (Natural_Capital_3/NC_Carrying_Capacity_3)))
Transfer_23 (Not in a sector)

OUTFLOWS:
NC__Harvest_3 = harvest_3
NC_Depreciation_3 = Natural_Capital_3*.02
NC_Carrying_Capacity_3 = 10000
NC_Growth_Rate_3 = 1

Harvest Rules:
harvest_units(t) = harvest_units(t − dt) + (harvest_growth) * dt

INIT harvest_units = 30

INFLOWS:
harvest_growth = if RULE = 3 then OPEN_ACCESS else
if RULE = 4 then SOLE_OWNER else
if RULE ≤ 2 then −(harvest_units − msy_har_units) else 0
The choice of level of harvesting units for open access (rule #3) and sole owner (rule #4) is determined here. For rules #1 and #2 harvest units are set here so that comparisons can be made with the other rules. Harvest units has no effect on harvests for rules #1 and #2.

CC% = .24
This is a constant harvest rate each year, with approximately 24% as the highest percent that the resource can sustain (at a growth rate of 1) without crashing. This percentage is then the maximum sustainable yield or MSY.

harvest = if RULE = 1 then HARVEST_LIMIT else if RULE = 2 then HARVEST_% else harvest_units*Harvest_Efficiency
HARVEST_% = stock_%*Total_Natural_Capital
This rule limits the harvest to the fraction stipulated (in stock_%) of the current stock of natural capital.

Harvest_Efficiency = Natural_Capital_1*.007
This equation gives the fraction of the current stock of natural capital harvested by each harvesting unit. For resources where there is no search problem (where harvest rate is not related to resource abundance) an absolute rate would be more appropriate.

HARVEST_LIMIT = CC%*NC_Carrying_Capacity_1
This equation sets the harvest limit at percent of carrying capacity set in CC%.

msy_har_units = harvest/Harvest_Efficiency
This variable simply calculates the harvest units that would be required to harvest either the HARVET LIMIT or the HARVEST %. This allows comparison of human capital with the sole owner and open access solutions.

RULE = 4
1 = HARVEST LIMIT as % of carrying capacity (.24 ~MSY @ growth = 1)
2 = harvest limit (% of current population)
3 = open access (total profits = 0)
4 = sole owner (total profits maximized)

stock_% = .285
Fraction of current population harvested. A fraction of about .285 comes close to maximizing the human capital from this policy.

Harvest Rules 2:

harvest_units_2(t) = harvest_units_2(t − dt) + (harvest_growth_2) * dt

INIT harvest_units_2 = 30

INFLOWS:
harvest_growth_2 = if RULE_2 = 3 then OPEN_ACCESS_2 else
if RULE_2 = 4 then SOLE_OWNER_2 else
if RULE_2 < = 2 then −(harvest_units_2-msy_har_units_2) else 0
Choice of level of harvesting units for open access (rule #3) and sole owner (rule #4) is determined here. For rules #1 and #2 harvest units are set here so that comparisons can be made with the other rules. Harvest units has no effect on harvests for rules #1 and #2.

CC%_2 = .24
Percent of carrying capacity harvested. This is a constant number each year. About 24% is the highest percent that the resource can sustain (at a growth rate of 1) without crashing = MSY.

HARVEST_%_2 = stock_%_2*Total_Natural_Capital
This equation limits the harvest to the fraction stipulated (in stock_%) of the current stock of natural capital.

harvest_2 = if RULE_2 = 1 then HARVEST_LIMIT_2 else if RULE_2 = 2 then HARVEST_%_2 else harvest_units_2*Harvest_Efficiency_2
Harvest_Efficiency_2 = Natural_Capital_2*.007
This gives the fraction of the current stock of natural capital harvested by each harvesting unit. For resources where there is no search problem (where harvest rate is not related to resource abundance) an absolute rate would be more appropriate.

HARVEST_LIMIT_2 = CC%_2*NC_Carrying_Capacity_2
Sets the harvest limit at percent of carrying capacity set in CC%.

harv_unit_trend_2 = TREND(harvest_units_2,5)*100

msy_har_units_2 = harvest_2/Harvest_Efficiency_2
This variable simply calculates the harvest units that would be required to harvest either the HARVEST LIMIT or the HARVEST %. This allows comparison of human capital with the sole owner and open access solutions.

profit_trend_2 = TREND(harvest_revenue_2-harvest_cost_2,5)
A trend is used rather than the most recent profit level because feedback from the biological sector can cause the most recent profit level to produce a false signal. A 5-year trend generates a somewhat slower response but tends to search out the maximum much more reliably.

RULE_2 = 4
1 = HARVEST LIMIT as % of carrying capacity (.24 ~MSY @ growth = 1)
2 = harvest limit (% of current population)
3 = open access (total profits = 0)
4 = sole owner (total profits maximized)

SOLE_OWNER_2 = if profit_trend_2 > = 0
then if harv_unit_trend_2 > .01 then effort_rate_2 else -effort_rate_2
else if harv_unit_trend_2 < = .01 then effort_rate_2 else -effort_rate_2
This is the profit maximizing solution. Simple search that assumes the relationship between profits and harvest units is a smooth single peaked function.

stock_%_2 = .285
Fraction of current population harvested. A fraction of about .285 comes close to maximizing the human capital from this policy.

effort_rate_2 = GRAPH(ABS(harvest_revenue_2-harvest_cost_
2-DELAY(harvest_revenue_2-harvest_cost_2,1)))
(0.00, 0.01), (1000, 0.12), (2000, 0.28), (3000, 0.46), (4000, 0.68), (5000, 1.00), (6000, 1.40), (7000, 1.84), (8000, 2.42), (9000, 3.20), (10000, 4.00)
Assumes the change in the rate of harvest effort reflects the magnitude of recent changes in the level of profits. If this response rate is set so that it is too sensitive, effort changes will be consistently too large in both directions (but with one dominating) and lead to a cascading increase or decrease in harvesting.

OPEN_ACCESS_2 = GRAPH(harvest_revenue_2-harvest_cost_2)
(−3000, −4.00), (−2400, −2.32), (−1800, −1), (−1200, −0.39), (−600, −0.12), (0.00, −1.11e-016), (600, 0.12), (1200, 0.3), (1800, 0.72), (2400, 2.08), (3000, 4.00)

Entry continues until positive profits are driven to zero. Exit occurs until negative profits are driven to zero.

Harvest Rules 3:

harvest_units_3(t) = harvest_units_3(t − dt) + (harvest_growth_3) * dt

INIT harvest_units_3 = 30

INFLOWS:

harvest_growth_3 = if RULE_3 = 3 then OPEN_ACCESS_3 else
if RULE_3 = 4 then SOLE_OWNER_3 else
if RULE_3 < = 2 then -(harvest_units_3-msy_har_units_3) else 0
Choice of level of harvesting units for open access (rule #3) and sole owner (rule #4) is determined here. For rules #1 and #2 harvest units are set here so that comparisons can be made with the other rules. Harvest units have no effect on harvests for rules #1 and #2.

CC%_3 = .24
Percent of carrying capacity harvested. This is a constant number each year. About 24% is the highest percent that the resource can sustain (at a growth rate of 1) without crashing = MSY.

HARVEST_%_3 = stock_%_3*Total_Natural_Capital
This equation limits harvest to the fraction stipulated (in stock_%) of the current stock of natural capital.

harvest_3 = if RULE_3 = 1 then HARVEST_LIMIT_3 else if RULE_3 = 2 then HARVEST_%_3 else harvest_units_3*Harvest_Efficiency_3
Harvest_Efficiency_3 = Natural_Capital_3*.007
This gives the fraction of the current stock of natural capital harvested by each harvesting unit. For resources where there is no search problem (where harvest rate is not related to resource abundance) an absolute rate would be more appropriate.

HARVEST_LIMIT_3 = CC%_3*NC_Carrying_Capacity_3
Sets the harvest limit at percent of carrying capacity set in CC%.

harv_unit_trend_3 = TREND(harvest_units_3,5)*100
See note for profit trend.

msy_har_units_3 = harvest_3/Harvest_Efficiency_3
This variable simply calculates the harvest units that would be required to harvest either the HARVEST LIMIT or the HARVEST percent. This allows comparison of human capital with the sole owner and open access solutions.

profit_trend_3 = TREND(harvest_revenue_3-harvest_cost_3,5)
A trend is used rather than the most recent profit level because feedback from the biological sector can cause the most recent profit level to produce a false signal. A 5-year trend generates a somewhat slower response but tends to search out the maximum much more reliably.

RULE_3 = 4
1 = HARVEST LIMIT as % of carrying capacity (.24 ~MSY @ growth = 1)
2 = harvest limit (% of current population)
3 = open access (total profits = 0)
4 = sole owner (total profits maximized)

SOLE_OWNER_3 = if profit_trend_3 > = 0
then if harv_unit_trend_3 > .01 then effort_rate_3 else -effort_rate_3
else if harv_unit_trend_3 < = .01 then effort_rate_3 else -effort_rate_3
This is the profit maximizing solution. Simple search that assumes the relationship between profits and harvest units is a smooth single peaked function.

stock_%_3 = .285
Fraction of current population harvested. A fraction of about .285 comes close to maximizing the human capital from this policy.

effort_rate_3 = GRAPH(ABS(harvest_revenue_3-harvest_cost_3-DELAY(harvest_revenue_3-harvest_cost_3,1)))
(0.00, 0.01), (1000, 0.12), (2000, 0.28), (3000, 0.46), (4000, 0.68), (5000, 1.00), (6000, 1.40), (7000, 1.84), (8000, 2.42), (9000, 3.20), (10000, 4.00)
Assumes the change in the rate of harvest effort reflects the magnitude of recent changes in the level of profits. If this response rate is set so that it is too sensitive, effort changes will be consistently too large in both directions (but with one dominating) and lead to a cascading increase or decrease in harvesting.

OPEN_ACCESS_3 = GRAPH(harvest_revenue_3-harvest_cost_3)
(−3000, −4.00), (−2400, −2.32), (−1800, −1), (−1200, −0.39), (−600, −0.12), (0.00, −1.11e-016), (600, 0.12), (1200, 0.3), (1800, 0.72), (2400, 2.08), (3000, 4.00)
Entry continues until positive profits are driven to zero. Exit occurs until negative profits are driven to zero.

Higher Level:

Higher_Level_Capital(t) = Higher_Level_Capital(t − dt) + (Accumulation − HLC_____Depreciation − HL_Services) * dt

INIT Higher_Level_Capital = 0

INFLOWS:
Accumulation = Tax_1 + Tax_2 + Tax_3

OUTFLOWS:
HLC____Depreciation = Higher_Level_Capital*.03
HL_Services = Control_Effort*Higher_Level_Capital
Control_Effort = 0.01
Share_or__Limit_2 = 0
Tax_rate_2 = 0.01
Tax_rate_3 = 0.01
Tax__rate_4 = 0.01

Human System 1:

Human_Made_Capital_1(t) = Human_Made_Capital_1(t − dt) + (HMC_Growth_Rate_1 − HMC___Depreciation_1 − Tax_1) * dt

INIT Human_Made_Capital_1 = 10
This is accumulated profits or wealth.

INFLOWS:
HMC_Growth_Rate_1 = harvest_revenue-harvest_cost

OUTFLOWS:
HMC___Depreciation_1 = Human_Made_Capital_1*.00
Tax_1 = Human_Made_Capital_1*Tax__rate_1
harvest_cost = harvest_units*unit_cost
harvest_price = 10
harvest_revenue = harvest_price*NC__Harvest_1
Tax__rate_1 = 0.00
unit_cost = 500

Human System 2:

Human_Made_Capital_2(t) = Human_Made_Capital_2(t − dt) + (HMC_Growth_Rate_2 − HMC___Depreciation_2 − Tax_2) * dt

INIT Human_Made_Capital_2 = 10
This is accumulated profits or wealth.

INFLOWS:
HMC_Growth_Rate_2 = harvest_revenue_2-harvest_cost_2

OUTFLOWS:
HMC___Depreciation_2 = Human_Made_Capital_2*.00
Tax_2 = Human_Made_Capital_2*Tax__rate_2

harvest_cost_2 = harvest_units_2*unit_cost_2
harvest_price_2 = 10
harvest_revenue_2 = harvest_price_2*NC__Harvest_2
Tax__rate_2 = 0.00
unit_cost_2 = 400

Human System 3:

Human_Made_Capital_3(t) = Human_Made_Capital_3(t − dt) + (HMC_Growth_Rate_3 − HMC___Depreciation_3 − Tax_3) * dt

INIT Human_Made_Capital_3 = 10
This is accumulated profits or wealth.

INFLOWS:
HMC_Growth_Rate_3 = harvest_revenue_3-harvest_cost_3

OUTFLOWS:
HMC___Depreciation_3 = Human_Made_Capital_3*.00
Tax_3 = Human_Made_Capital_3*Tax__rate_3
harvest_cost_3 = harvest_units_3*unit_cost_3
harvest_price_3 = 10
harvest_revenue_3 = harvest_price_3*NC__Harvest_3
Tax__rate_3 = 0.00
unit_cost_3 = 300

Profit-based Decisions
harv_unit_trend = TREND(harvest_units,5)*100

profit_trend = TREND(harvest_revenue-harvest_cost,5)
A trend is used rather than the most recent profit level because feedback from the biological sector can cause the most recent profit level to produce a false signal. A 5-year trend generates a somewhat slower response but tends to search out the maximum much more reliably.

SOLE_OWNER = if profit_trend > = 0
then if harv_unit_trend > .01 then effort_rate else -effort_rate
else if harv_unit_trend < = .01 then effort_rate else -effort_rate
This is the profit maximizing solution. Simple search that assumes the relationship between profits and harvest units is a smooth single-peaked function.

effort_rate = GRAPH(ABS(harvest_revenue-harvest_cost-DELAY(harvest_revenue-harvest_cost,1)))
(0.00, 0.01), (1000, 0.12), (2000, 0.28), (3000, 0.46), (4000, 0.68), (5000, 1.00), (6000, 1.40), (7000, 1.84), (8000, 2.42), (9000, 3.20), (10000, 4.00)

Assumes the change in the rate of harvest effort reflects the magnitude of recent changes in the level of profits. If this response rate is set so that it is too sensitive, effort changes will be consistently too large in both directions (but with one dominating) and lead to a cascading increase or decrease in harvesting.

OPEN_ACCESS = GRAPH(harvest_revenue-harvest_cost)
(−3000, −4.00), (−2400, −2.32), (−1800, −1), (−1200, −0.39), (−600, −0.12), (0.00, −1.11e-016), (600, 0.12), (1200, 0.3), (1800, 0.72), (2400, 2.08), (3000, 4.00)
Entry continues until positive profits are driven to zero. Exit occurs until negative profits are driven to zero.

Sector 1:

CV_of_HMC = SQRT((.5*((Human_Made_Capital_1-Mean_HMC)^ 2 + (Human_Made_Capital_2-Mean_HMC) 2 + (Human_Made_Capital_3-Mean _HMC) 2)))/Mean_HMC
Mean_HMC = Total_Human_Made_Capital/3
Total_All_Capital = Higher_Level_Capital + Total_Human_Made_Capital + Total_Natural_Capital
Total_Human_Made_Capital = Human_Made_Capital_1 + Human_Made_Capital_2 + Human_Made_Capital_3
Total_Natural_Capital = Value_of_NC*(Natural_Capital_1 + Natural_Capital_2 + Natural_Capital_3)
Value_of_NC = 10
Value of Natural Capital in $ per unit

Not in a sector:

Transfer_12=Adjusted_Trans_Rate_12*(Natural_Capital_1-Natural_Capital_2)

OUTFLOW FROM: Natural_Capital_1 (IN SECTOR: Ecosystem 1)

INFLOW TO: Natural_Capital_2 (IN SECTOR: Ecosystem 2)
Transfer_23=Adjusted_Trans_Rate_23*(Natural_Capital_2-Natural_Capital_3)

OUTFLOW FROM: Natural_Capital_2 (IN SECTOR: Ecosystem 2)

INFLOW TO: Natural_Capital_3 (IN SECTOR: Ecosystem 3)
Adjusted_Trans_Rate_12 = if Share_or__Limit = 1 then Nat_Trans_Rate_12 − (Nat_Trans_Rate_12 * (HL_Services/(50 + HL_Services))) else Nat_Trans_Rate_12 + ((1 − Nat_Trans_Rate_12) * (HL_Services/(50 + HL_Services)))
Adjusted_Trans_Rate_23 = if Share_or__Limit = 1 then Nat_Trans_Rate_23 − (Nat_Trans_Rate_23 * (HL_Services/(50 + HL_Services)))

```
else Nat_Trans_Rate_23 + ((1 - Nat_Trans_Rate_23) * (HL_Services/
(50 + HL_Services)))
External_Influences_1 = RANDOM(1,1)
External_Influences_2 = RANDOM(1,1)
External_Influences_3 = RANDOM(1,1)
Nat_Trans_Rate_12 = .1
Nat_Trans_Rate_23 = .1
Share_or__Limit = 1
```

chapter four

Scale misperceptions and the spatial dynamics of a social–ecological system

James Wilson, Robert Costanza, Bobbi S. Low, and Elinor Ostrom

Contents

The interactions between an ecosystem and the human rules for the use of that system can be very complex. This complexity makes it hard to design foolproof and sensible rules. Here we explore a particular set of difficult questions: What are the consequences of misunderstanding or misperceiving the structure of populations we wish to exploit? What if the "scale" of natural populations and their interactions do not match the scale of our decisions? What if we think we are managing a single large population, when in fact there are multiple, small, spatially discrete populations?

These are important and relevant questions. In the 1950's and 1960's, many environmental programs were initiated at the national and international level. As a consequence, both the theory and practice of environmental and resource management have focused on a scale of authority appropriate for national and international regulatory bodies. In fisheries, for example, the first serious attempts at management began with the international organizations for the northwest and northeast Atlantic and the whale and the tuna

1-56670-389-1/01/$0.00+$.50

commissions. With the advent of extended fisheries jurisdiction (i.e., the 200-mile limit), national organizations took over much of the authority of the international bodies, but often retained the same scale of regulation (generally over large areas involving thousands of square kilometers). Of necessity, regulatory bodies operating at this scale are forced to ignore the fine-scale aspects of the systems they regulate.

The poor performance of regulated ocean fisheries provides ample reason to question the scale of regulatory attention. A number of recent papers (Hutchins and Meyers, 1995; Ames, 1996; Wilson et al., 1998; Sinclair, 1988) have focused attention on the existence of populations at a smaller scale than that usually managed by national or international regulatory authorities. The usual thrust of these arguments is that regulatory regimes that ignore smaller scale events and phenomena such as habitat and local stocks may lead, inadvertently, to the erosion of the spatial structure of a population and the depletion of the resource. Put differently, scale misperceptions might lead to a different form of overfishing than that usually hypothesized. In particular, rather than overfishing simply by harvesting too many fish, it may be possible to overfish by inadvertently destroying the spatial structure of a population.

The existence of localized spawning groups for a number of important species has been known for a long time (Sinclair, 1988). But are these local-scale spawning groups relatively distinct populations, together forming the "structure" of the larger fishery of which they are a part? If so, they may need to be managed separately. Or, are these groups simply the spatial expression of a larger population, in which case it may be appropriate to ignore the local particulars? The principal question is really: At what scale (or scales) should fishery management operate?

Many marine biologists have argued that localized populations (and thus issues of scale) are irrelevant to management because of the high rates of larval mixing among marine populations (Forgarty et al., 1997). If a local population is extinguished, it is likely that its population space will be quickly recolonized by members of other populations. In this view, because the population is panmictic, there is no need to manage local populations separately. From the management perspective, only the aggregate population is relevant for the application of restraints or rules. If this view of population behavior is correct, then the scale misperceptions with which we are concerned would appear to have no practical impact.

An alternative view, the metapopulation perspective, is more common among terrestrial ecologists (Gilpin, 1996). In metapopulation theory, a local population is relatively discrete and reproductively separated to some degree from other local populations. The reasons for the separation might be genetic, imprinted, or learned behavior that brings members of the local population back to the same spawning site. However, fidelity to spawning grounds is imperfect and a few members of any population may well stray to other populations. If local populations are extinguished due to natural or man-made causes, strays from other proximate populations can wander to the spawning site and recolonize the population "space," although more slowly than in panmictic situations. If local extinction is rare, recolonization can restore the spatial structure of the metapopulation and maintain its resiliency in the face of local extinctions. However, if the extinction rate of local populations exceeds the recolonization rate, the resilience of the metapopulation is eroded.

From a management perspective, a metapopulation differs from a panmictic population principally in terms of (1) the causes or patterns by which fishing might bring about the collapse of a large population; (2) the speed of recolonization; and (3) the mechanisms by which local and large populations can, or are likely to, rebuild on their own or through human intervention.

All of these differences are critical to appropriate strategies for the management of fisheries. A panmictic population, for example, is reduced simply by taking too many fish from the entire (aggregate) population; only the total take, not its spatial distribution, matters. Rebuilding a panmictic population should depend principally upon the normal spawning and recruitment processes of the population and, as a result, can occur quite rapidly so long as fishing leaves an adequate spawning biomass. The pattern of collapse of a metapopulation through fishing can be best summarized as "piece by piece" disappearance until the overall population structure is reduced to fragmented remnant local populations. A metapopulation may take a long time to recover from over fishing and local population extinction, depending upon the factors governing recolonization (especially how potential inmigrant members of the population acquire the behavior that leads to spawning site fidelity). Management strategies for rebuilding metapopulations may depend upon knowledge of behavioral and other aspects of a species' life history, factors about which we now have little firm knowledge.

While answers to these questions are critical to the design of appropriate management regimes, it is almost impossible to obtain sufficient empirical data to test these hypotheses. One can, however, build models of panmictic and metapopulations to explore these questions. Consequently, here we explore a series of illustrative models in which local populations—modeled as either panmictic or metapopulation structures—are managed as if they composed a single large population. These models are a dynamic version of the generic bioeconomic model of a single stock (Clark, 1976; Anderson, 1986), and are used to investigate the circumstances under which common regulatory procedures might lead to depletion of the fishery.

4.1 The model

The basic model used here is an extension of the one used in the previous chapter. The principal differences between this and our earlier model are (1) the populations are arranged in an implicit "triangular" spatial structure that allows fish from any of the three populations to move directly to either adjacent population, and (2) we examine the different ways these populations might interact—i.e., as a panmictic population or as a metapopulation—under various rules that managers might use in the fishery (i.e., open access, constant percent quota, and sole ownership).

In each version of the model the three "local" populations are given identical carrying capacities. As is common in fisheries management (Sinclair, 1988), the regulatory authority perceives or treats the three local populations as if they were a single unified population and manages accordingly. Typically, this management approach is based on (1) the assumption that populations in the ocean have a high level of mixing and consequently act as if they were single populations, and/or (2) the often high costs and difficulty of monitoring and assessing separate populations (Forgarty et al., 1997).

The model implements the three management rules in ways that tend to dampen feedback from their (self) implementation, i.e., the rules are implemented so that they do not, by themselves, tend to destabilize the system. The open access rule is constructed so that when average profits per harvester are positive, entry takes place; when average profits per harvester are negative, exit takes place; obviously, when profits are zero, no entry or exit takes place. Furthermore, the entry (or exit) response to non-zero profits increases nonlinearly as profits diverge farther from zero. This tends to prevent overshoot and oscillations due to too many or too few boats relative to zero profits. This formulation implicitly assumes boats and operators entering and exiting the fishery have no problem finding alternative employment, and that there are no regulatory barriers that impede inward mobility. Additionally, it is assumed there are no lags that might cause entry and exit to continually over or undershoot the level of effort associated with zero average profits. We considered, but rejected, a rule that based entry and exit upon a trend in average profits rather than just current average profits; this would have led to greater stability in the model, but it violates the basic strategy behind open access entry and exit (i.e., move before your competitor does).

The constant percentage quota rule is implemented as a simple translation of the number of fish to be caught (constant percent times stock size) into the right number of boats to catch that number. Here also there are assumed to be no barriers or lags that might impede the implementation of this rule or contribute to a problem of overshoot or undershoot. In particular, it is assumed that measurements of the current size of the stock(s) are without error and are analyzed correctly with the resulting quotas implemented in a timely fashion.

The profit maximizing sole owner rule is implemented as a search process that compares "the owner's" past actions (adding vs. subtracting boats) with the subsequent results (more or less profit). The rule uses a 5-year trend to allow the impact of more or fewer boats to work its way through all the fisheries so that the sole owner can sort out signal from noise in circumstances when populations are variable. This way of implementing the profit maximizing rule tends to slow the rate at which maximum profits are achieved, but has the advantage of being much more robust (in the sense of finding the true maximum rather than some local maximum) and stable in circumstances of high population variability.

All three of these rules could have been designed with alternative formulations that might be argued to be more realistic, e.g., "sticky" entry and exit, errors and delays in measurement, and so on. However, our objective here is to isolate problems that arise because of a misperception of scale. Consequently, we have tried to minimize the kinds of dynamic problems that might arise if these management rules were more "realistic" so that we might better recognize any scale-related problems.

The two population types are differentiated from one another in a very simple way. In panmictic populations no critical minimum population size is specified but for the metapopulation versions a critical minimum size was specified for each of the local populations. This was always set at .05 of carrying capacity, K. This size assumes that below the critical minimum, there are too few individuals in the local population either to attract conspecifics or to spawn successfully. When that critical minimum size is reached, both recruitment and in-migration cease and the local population is eventually extinguished. The model does not contain a mechanism for recolonization after extinction occurs. This treatment thus mimics short-term metapopulation dynamics in which extirpated local populations are rarely recolonized, and illustrates nicely the effect of local extinctions.

The long-run dynamic of each local population is characterized by a Schaefer stock/recruitment relationship:

$$S_{t+1} = rS_t(1 - S_t / K) \tag{4.1}$$

where S_t is the population numbers at time t, r is the intrinsic growth rate of the population, and K is the carrying capacity stated in terms of the maximum number of individuals in the population (see Figure 4.1). Taken by itself, the equation states that the population in the current period is a function of the size of the population in the previous period. At low population levels the population experiences rapid growth, but at high population levels the rate of growth declines until it reaches zero at a population level corresponding with carrying capacity. At its carrying capacity, K, there is no tendency to change and the population is in a stable equilibrium. As mentioned above, in the metapopulation version, the recruitment relationship is modified with a requirement for a critical minimum population size.

In addition to recruitment, each local population is affected by movements of fish to and from adjacent populations and by withdrawals due to harvesting. The basis for movement between populations is defined as:

$$\text{Trans}_{a,b} = \text{transrate} * (S_a/K_a - S_b/K_b) \tag{4.2}$$

where the amount of transfer or migration between populations a and b is given by a transfer rate, "transrate," which is a function of the difference in the density (relative to carrying capacity) of the two populations. At a value for "transrate" of 0.5, fish achieve a free distribution within one period; at a value of 0.0, fish cannot move between populations.[1]

[1] At values above 0.5 the movement of fish tends to overcompensate for density differences among the populations and leads to large cyclical swings. We limit tests of the model to values at or below 0.5.

The harvest rate for each local population is determined by the effort allocated to that population. In each period of the simulation, changes in the total level of harvesting effort (i.e., for all three populations) are determined by the centralized management authority using one of the three harvest rules described above. Total effort is then allocated to each population by a mechanism that allocates the sum of (1) changes in total fishing effort, and (2) a specified portion of existing effort, given by "switchrate," so that profits per boat are equalized in each of the three local fisheries. In other words, the management rule determines the overall level of effort for all three fisheries taken together, but the allocation of effort to each of the three fisheries is driven by the relative profits earned by boats fishing each local population. When the value of the variable "switchrate" is equal to 0.5, fishermen achieve a free distribution quickly; basically, fishermen are free to move boats and equipment and rapidly equalize profits in the three local fisheries. When "switchrate" equals zero, boats are confined to the fishery to which they were initially assigned and profits are not equalized except by chance. This formulation of the model assumes that fishermen are able to perceive the small-scale population structure that managers either fail to perceive or choose to ignore.

The third factor determining harvest rates is the efficiency of harvest, where efficiency is measured in terms of percent of the population harvested annually by a single boat. Efficiency is assumed to be partly due to the number of other boats in the fishery (the more boats the less efficient each boat) and the behavioral characteristics of the fish in the populations which alters search costs, especially as populations decline. Finally, harvest rates for each of the three local populations are also partly determined by the differences in the costs of harvesting each local population, i.e., differences in costs lead to differences in profits and eventually to differences in the allocation of effort to each local population.

Given this description of the model, we return now to the broad question we wish to ask: Does management misperception of the appropriate population scale make a difference? And, if so, under what circumstances? Our originating premise was that the underlying population structure—whether of panmictic or metapopulation type—is likely to be an important differentiating circumstance. Also, given the way we have articulated the problem, our intuition leads us to expect scale misperception problems when circumstances drive one or all local populations towards the critical minimum. Broadly, these circumstances might be expected to occur when (1) "deterministic" aspects of the model (e.g., an open access rule) would lead to low population levels, and (2) when external or internal sources of variability also contribute to low population levels. Following are several hypotheses about when those circumstances might arise:

1. An open access rule is known from analytical models to lead to depleted populations given sufficient market demand. It would seem

reasonable, therefore, that an open access rule would frequently drive populations into a range near the critical minimum and that metapopulation structures would be especially vulnerable to this management rule. On the other hand, analytical models indicate that both the constant percent quota rule and the sole owner rule maintain populations at relatively high levels well removed from the critical minimum level. Consequently, one would not expect these two rules to lead to scale misperception problems.

2. The ability of fishermen to switch easily between populations should reduce the profit differentials between populations and the tendency of a local population to be depleted. In other words, as a local population approached depletion or extinction, profits should fall and cause effort to move to healthier local populations, thereby providing some measure of protection against management misperception of the appropriate scale.
3. The extent of migration or transfer of fish from one population to another should affect the viability of local populations; in particular, we expect that high levels of movements of fish between populations would tend to make the three local populations act like a single large population. At low rates of interpopulation movement, one might expect individual populations to be more vulnerable to depletion or extinction.
4. Differences in the cost of exploiting one local population relative to another might create an economic preference that puts strong pressures on, and might possibly deplete or extinguish, the least costly-to-exploit population.
5. Finally, we would expect variability, whether generated as part of the internal dynamics of the model or from external sources, to increase the possibility of populations reaching critical minimum levels.

To test these expectations we initiated an extensive series of sensitivity runs that covered a reasonable sub-set of the values of the relevant variables. The general method behind our explorations was to start with a set of parameter values, or circumstances that generate a dynamic solution that mirrors or approximates the results expected from steady-state-analytical models (Clark, 1976). Due to the complexity of the model, a steady-state "base" of this sort helps greatly in the identification and analysis of more complex dynamic behaviors.

Our first test was to compare our three-population panmictic model with a "true" single population model. Our purpose was to see whether the migration of fish and fishermen in our panmictic model led to different results than might be found with a "true" single population. We set the maximum migration rates for fish and fishermen to 0.25 (i.e., up to 25% of each population could conceivably move each year). Both models produce similar time paths for each of the three management rules (Figure 4.1); however, the numerical results differ significantly.

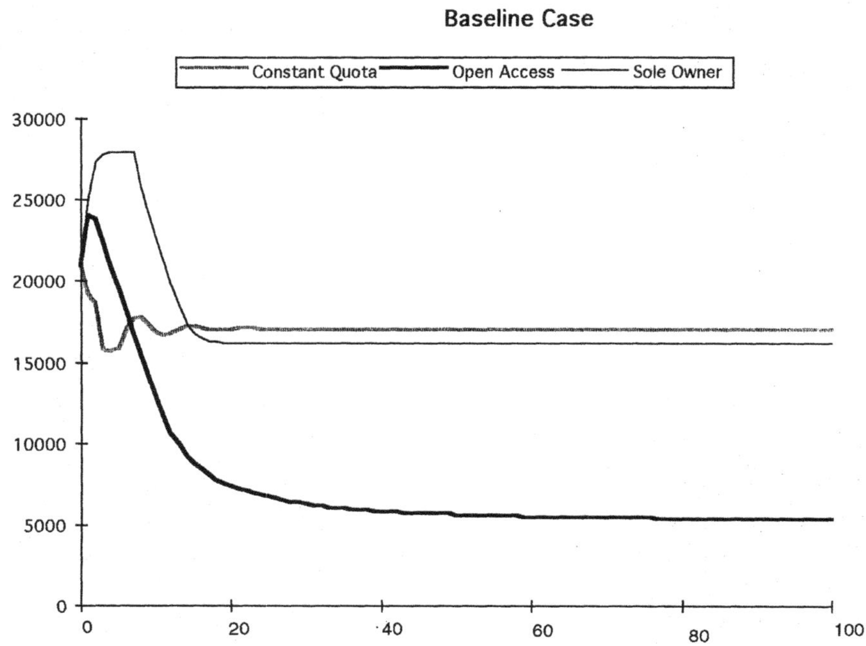

Figure 4.1 Baseline case: sum of all populations for three management rules.

Comparing Tables 4.1 and 4.2, it is readily apparent that all three rules tend to result in lower stock sizes when applied as aggregate rules to the three-population case. The reason for this is that the migration of fish and fishermen from population to population tends to mask the true state of the overall population. The migration of fish is always towards the smaller population. In the case of healthy populations—such as under the sole owner and constant quota rules—the smaller populations generally tend to be the populations with a higher growth rate. Consequently, migration tends to raise the average growth rate of the system, leading to higher sustained harvests but at lower sustained population levels. This effect is interesting, but is totally dependent upon the assumed parabolic shape of the recruitment curve and so might be legitimately argued to be simply an artifact of the model; there are few if any marine fish populations for which a recruitment curve has been validated over the full range of population size, much less one with a parabolic shape (Hall, 1988). The movement of fishermen from population to population, on the other hand, leads to the preponderance of harvests coming from the largest or healthiest sub-populations. That is, as the component populations vary in size, fishermen always move to the largest populations; this increases their profits, but also contributes to higher sustained average catches at lower average sustained population levels (compare Tables 4.1 and 4.2).

Table 4.1 Single Population Model

decision rule	open access	constant % quota	sole owner
avg. harvest as % of carrying capacity	11	23	22
avg. pop size as % of carrying capacity	18	61	59

Table 4.2 Panmictic Population with Three Local Populations

decision rule	open access	constant % quota	sole owner
avg. harvest as % of carrying capacity	08	15	14
avg. pop size as % of carrying capacity	08	57	54

In the case of the open access rule, however, the growth effects of migration work in the opposite direction (Figure 4.2). Here migration tends to move towards stocks with lower growth, leading to significantly lower sustained harvests than might be expected to occur in a true panmictic population. In this instance, more significance might be attached to the result since it does appear that at very low population levels some sort of recruitment relationship exists (Hutchins and Meyers, 1995). In other words, given the assumed migration of fish from stronger to weaker stocks and of fishermen from weaker to stronger stocks, management's misperception of a single large stock might lead to an understatement of the severity of the overfishing problem.

From this baseline we began to explore the model for sources of internal dynamic variability. Two variables unique to the three-population structure, i.e., the rate at which fish migrate between populations—"transrate"—and the rate at which fishermen switch between populations—"switchrate"—are important sources of internal variability. Sensitivity analyses for all combinations of values of "transrate" and "switchrate" of 0.0, 0.1, . . . 0.5 were run. These runs point to an interesting internal dynamic in the model, namely, as fish and fishermen attempt to adjust at varying rates towards a "free distribution" or "equal profits," there tends to be some overshoot since movement

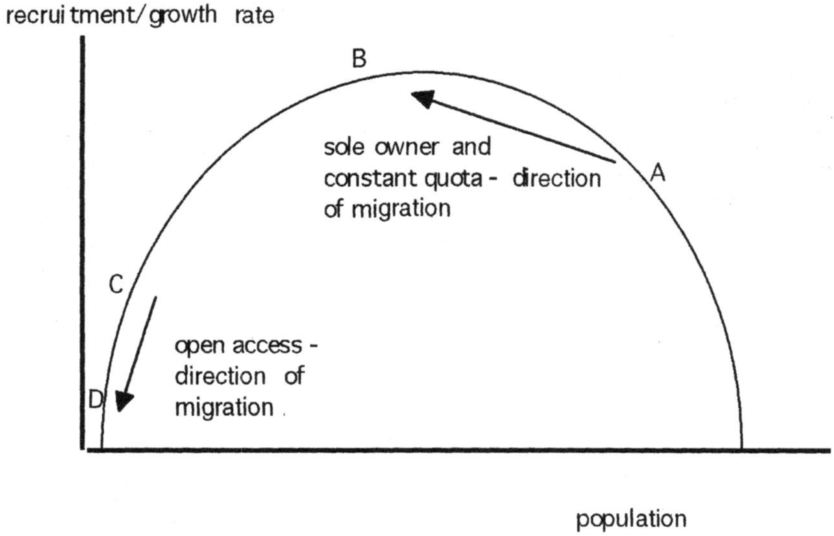

Figure 4.2 Population growth and migration recruitment/growth rate.

from two populations towards one is uncoordinated. Figure 4.3 shows how this internal dynamic affects the three populations. In this instance, the results were for an open access rule in which fish were restricted to their originating population, but up to 25% of fishermen were free to move between populations each year. Similar results arise when fish are free to move between populations. Clearly, this aspect of the model points to a mechanism that might contribute to populations approaching the critical minimum.

Other sources of internal dynamic variability (not illustrated here) include:

1. The intrinsic growth rate of the population, r. In the steady-state version, r is set at 1.0 for all three local populations. At a value for r in the vicinity of 2.00 large periodic fluctuations in the local populations begin to occur. At values around 4.0 chaotic fluctuations start (the exact value at which periodic and chaotic fluctuations occur depends upon the amount of harvesting, May, 1974). In all the examples that follow a value of $r = 1$ was used for all three local populations.
2. The reaction time and rates of response of decision makers are normally set so they will yield a steady-state solution. However, when delays in the receipt or analysis of information, or when there are errors of measurement or when a host of other factors are incorporated in the model, all three decision (or management) rules consistently tend to overshoot or undershoot their effort targets, leading to circumstances of high variability. Again, in all the examples that

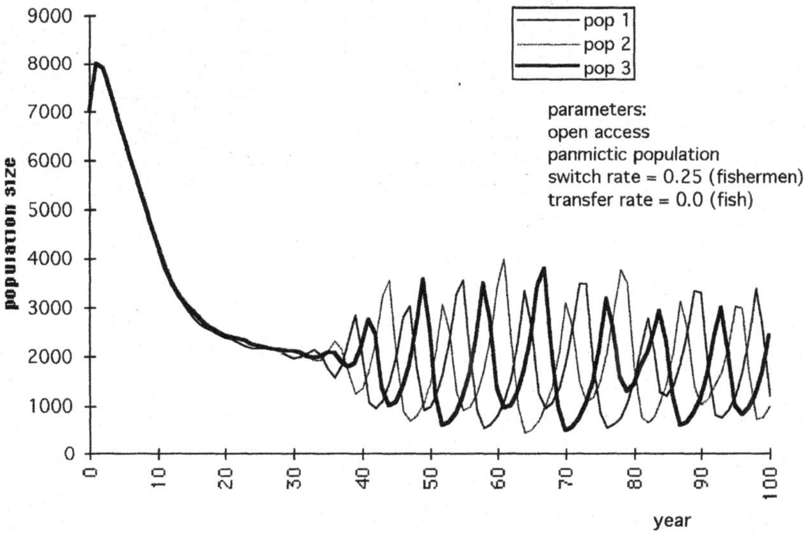

Figure 4.3 Periodicity in populations arising from migration of fishermen.

follow, model formulations consistent with a steady-state solution were used in order to exclude these sources of variability.

Additionally, one would expect a fishery of this sort to be subject to a variety of external sources of variability—weather, human intervention, etc. If all external and internal sources of variability were included in the model, it would yield results that are almost immune to analysis. Consequently, in the descriptions of the model from this point on, we restrict the variability in the model to two sources: variability due to interpopulation movements by fish and fishermen, and external variability that affects only recruitment to the population.

4.2 Testing hypotheses

Hypothesis 1: Does the management rule make a difference? The answer to this question is clearly yes—but only if the system is subject to some (internal or external source of) variability. To illustrate the importance of variability we re-ran the baseline model with levels of externally induced recruitment variability ranging from 0 to 200 percent. (This is not as high as it might seem. In a typical run, 200 percent recruitment variability results in average population variability from year to year in the range of 20 to 25%. The figure is greater for the open access case, in which the fishery is heavily dependent upon recruitment, and less in the sole owner and constant quota cases where high standing populations provide more of a buffer.) In the panmictic case with the open access rule (not shown) local populations frequently fall to very low levels and then rebuild only very slowly. In the metapopulation case with the open access rule, extinction of populations begins to occur with recruitment variability as low as 10% (see Figure 4.4a). Here the model leads us to conclude that results of open access are likely to be worse than expected when the panmictic case is compared with a single population case, and far worse in the metapopulation case.

The other two management rules tend to be more robust in the face of recruitment variability, but they also begin to yield extinctions in the metapopulation case when recruitment variability reaches 60–100% (Figures 4.4b and 4.4c). Since these are rather low levels of variation for marine systems, the model strongly suggests that rules that are optimal for single populations may lead to extinctions of local populations (that may be interpreted as the depletion of a single large population) when there is a scale misperception problem.

Hypotheses 2 and 3: Does the migration of either fish or fishermen tend to protect local populations against depletion? In both instances the answer is yes, up to a point. If, by chance, one of the local populations declines, migration of fish will tend to reinforce that population and at the same time fishermen—attempting to equalize profits—will desert the population and move to others where densities and profits are higher. Consequently, the

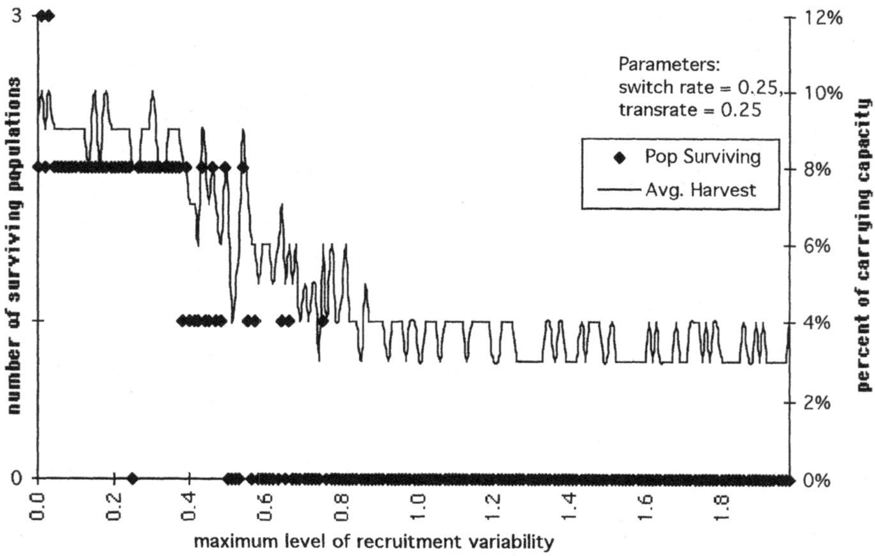

Figure 4.4a Population extinction and harvest: open access with metapopulations.

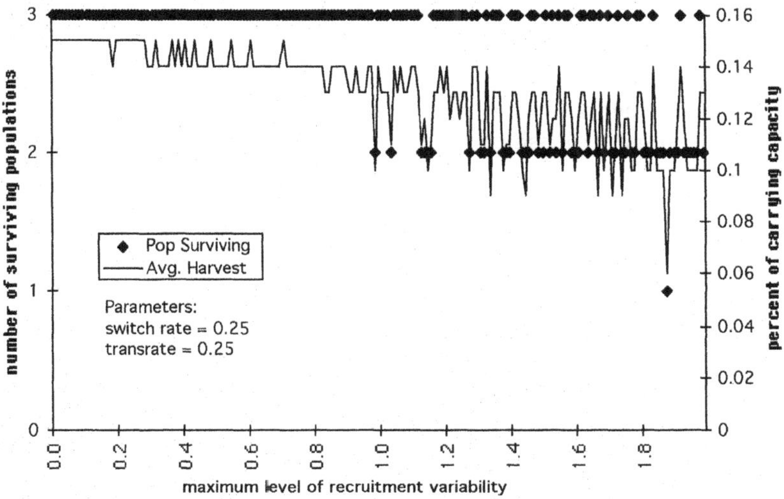

Figure 4.4b Population extinction and harvest: constant quota with metapopulations.

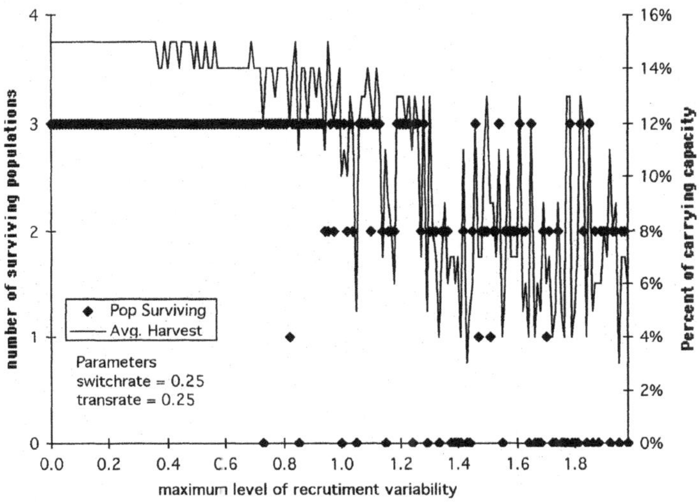

Figure 4.4c Population extinction and harvest: sole owner rule with metapopulations.

migration of both fish and fishermen tends to stabilize, or protect, local populations. The extent to which this is true depends upon the rate at which migration can take place. At very low permitted rates the protection effect is very weak. So long as there is little or no variability in the local populations this is not a problem; however, with almost any level of variability present, local populations tend to be more vulnerable to extinction than they are with moderate permitted migration of either fish or fishermen. At high permitted rates, the migration of either fish or fishermen can have a strongly destabilizing effect. Consider a situation in which fish are perfectly mobile and differences in the density of fish occur between local populations. Assume two populations, a and b, are of equal density and the third, c, is of lower density.

Table 4.3a **Population Size as a Percent of Carrying Capacity, (Open Access, Metapopulation, with Recruitment Variability = 0)[1]**

	Switchrate (boats)					
Transrate (fish)	**0.0**	**0.1**	**0.2**	**0.3**	**0.4**	**0.5**
0.0	0.18	0.17	0.18	0.12*	0.09**	0.00***
0.1	0.18	0.18	0.17	0.17	0.05**	0.00***
0.2	0.18	0.18	0.17	0.17	0.12*	0.00***
0.3	0.18	0.18	0.11*	0.13*	0.00***	0.00***
0.4	0.18	0.18	000.17	0.13*	0.12*	0.00***
0.5	0.18	0.17	0.17	0.11*	0.08**	0.00***

[1]*The number in each element of the table gives the average population as a percent of carrying capacity. *'s indicate the number of extinguished populations.*

Populations a and b respond independently to this difference; if the "permitted" migration rate (transrate) is high enough, both populations would tend to send enough migrants to c to erase the density differential. The result is twice as many fish moving to c as would be necessary to actually equalize densities and, consequently, large periodic swings in all three populations would begin. The same is true for fishermen migrating in response to unequal profits. From the human perspective this kind of destabilizing behavior is a watery equivalent of the hog cycle; it is not hard to imagine fishermen outsmarting themselves by collectively overreacting. Whether fish are this smart or not is hard to say. Whatever the case, the model does suggest that there are circumstances where this kind of behavior could destabilize a fishery that management incorrectly perceived to be a single large population.

Hypothesis 4: Do differences in the cost of exploiting one local population relative to another make a difference? The behavior of the model

Table 4.3b **Population Size as a Percent of Carrying Capacity, (Open Access, Metapopulation, with Recruitment Variability = 0.25)[2]**

	Switchrate (boats)					
Transrate (fish)	**0.0**	**0.1**	**0.2**	**0.3**	**0.4**	**0.5**
0.0	0.18	0.17	0.18	0.12*	0.09**	0.00***
0.1	0.16	0.12	0.10	0.09**	0.00***	0.00***
0.2	0.14*	0.11*	0.09**	0.18	0.00***	0.00***
0.3	0.14*	0.08**	0.18	0.00***	0.00***	0.00***
0.4	0.13*	0.15*	0.08**	0.13*	0.00***	0.00***
0.5	0.14*	0.04**	0.13*	0.00***	0.00***	0.00***

[2]*The number in each element of the table gives the average population as a percent of carrying capacity. *'s indicate the number of extinguished populations.*

indicates that this is likely to be much less of a problem than we had anticipated. The reason, very simply, is that the migratory behavior of both fish and fishermen (at moderate levels, i.e., 'transrate' and' switchrate' < 0.10–0.25) tends to protect populations that might otherwise be subject to heavy pressures. For example, if a population is fished heavily relative to others, its density and numbers fall, then fish from other populations tend to be attracted via migration and fishermen tend to be repelled by falling profits. Of course, as pointed out above, migration of both fish and fishermen can overcompensate, destabilize the fishery, and lead to a greater chance of extinction.

4.3 *Conclusions*

Our interest in scale misperceptions arises from the historical events that have led to the large-scale management of fisheries when, at the same time, there appears to be strong evidence that spawning for many fishery populations is relatively localized. This misperception of the appropriate ecological scale, on its face, could possibly lead to serious management problems. Consequently, we reformulated the basic model we have been using so that it was capable of investigating the implications of scale misperceptions. Our intention was to use the model to put a little more logical "meat" on our intuition that these misperceptions might lead to management problems.

What we found was that the extent of the problem depends greatly upon the kind of population structure assumed for the observed localized

Table 4.4 **The principal variables for each version of the model, their default or normal values and the range of values for which sensitivity runs were conducted.**

Variables**	Steady state values	Panmictic Population Normal Value	Panmictic Population Sensitivity Range	Metapopulation Normal Value	Metapopulation Sensitivity Range
transfer rate	0.00	0.25	0–0.5	0.25	0–0.5
switch rate	0.00	0.25	0–0.5	0.25	0–0.5
cost differences	0.00	0.00	0–100	0.00	0–100
critical minimum	0.00	0.00	n.a.	0.05	n.a.
recruitment variability	0.00	0.00	0–200	0.00	0–200

***Transfer rate—'transrate'—gives the maximum proportion of fish that can migrate from one population to another. Switch rate—'switchrate'—gives the maximum proportion of boats that can switch from one to another population. Cost differences—'cost diff'—gives the percent differences in the costs of fishing each of the three local populations. Recruitment variability—'vary'—gives the maximum percent random variation around the calculated level of recruitment.*

spawning groups. When local populations are modeled as if they were a panmictic population, that is, one that mixes freely and uses local spawning areas in some proportion to its overall state, scale misperceptions tend to lead to fewer management problems. The model suggests that there might be a tendency for standard management approaches to understate the extent of overfishing at low population levels, but this conclusion is weak, at best.

On the other hand, when local populations were modeled as a metapopulation, that is, one in which local spawning groups are relatively independent of one another, our results suggest management misperceptions of appropriate scale might be a serious cause of overfishing. In particular, under conditions of high variability, which are very common in marine populations, all three management rules tended to lead to the "piece by piece" reduction of the overall metapopulation. The open access rule, which we used as a worst case comparison, leads to the quick extirpation of localized stocks. The constant percent of stock quota and the sole owner rules, which are generally considered to be "optimal," proved to be much more robust than the open access rule, but also led to the "piece by piece" reduction of the overall metapopulation even with moderate levels of fishing effort and population variability.

These conclusions direct our attention to another way that overfishing can occur. Conventional wisdom and the standard scientific view of overfishing involves catching so many fish that a population cannot sustain itself. Our model suggests that overfishing can occur when we misperceive the appropriate scale at which populations operate. Under these circumstances, what are thought to be optimal rules can lead to the destruction of local substocks even though only moderate levels of fishing effort are employed. The further implication is that the avoidance of overfishing may involve much more than simply catching only the "right number" of fish. We may need to pay attention to the structure of local populations and the habitat and other biotic and abiotic factors necessary for their continued existence. This implies an emphasis on where, when, and how fishing takes place rather than simply an emphasis on the "right number" to catch. From a social perspective, the model strongly suggests a need to move away from our current emphasis on centralized management organizations, and to cultivate the growth of complementary local management organizations that can deal effectively with local ecological phenomena.

chapter five

Sustainability, equity, and efficiency of irrigation infrastructures

Nirmal Sengupta, Swati Sheladia, and Elinor Ostrom

Contents

1-56670-389-1/01/$0.00+$.50

In the cases explored in Chapters 3 and 4, human actors inherit and exploit an ecosystem where the stock of natural capital has an initial value, and the flow of units is primarily affected by natural processes and human harvesting activities. The major focus of the rules examined in those systems is related to restrictions placed on harvesting activities. In this chapter, however, we focus on how the use of some ecosystems requires human actors to design, construct, and operate physical capital that changes the relationships among the stocks and flows of the natural system. In the development of irrigation systems, for example, individuals invest in planning engineering works that divert water from its natural courses, store water for future use, or extract water from subsoil deposits. After these major works are constructed, farmers must also construct channels that distribute the water to secondary channels or directly to the farmers' fields. Quite often it is necessary to level the land in order to distribute water uniformly over agricultural fields. These costly investments in physical infrastructure change the structure and operation of the biophysical system itself.

Achieving the long-term benefits of these major investments in physical capital depends on finding ways to operate and maintain the system year after year. To do this, someone—either a government agency or the farmers themselves—must invest time and energy in designing rules that allocate water to those who will be using it, determine how they should contribute to the cost of operational and maintenance activities, and create the incentives to insure that individuals actually follow these allocation and maintenance rules. Otherwise, natural processes lead inevitably to the deterioration of the physical works and eventually to their breakdown and collapse.

Investing in irrigation infrastructure involves two types of activities and time frames. The first investment is in the design and construction of engineering works, and the principles are well understood. Many talented people are attracted to the study and practice of engineering major irrigation systems. In deciding whether a system should be built or not, engineers use standard methods for calculating annual benefits and costs. From these data, they ascertain the expected rate of return (for a set time horizon such as 25 or 50 years) using a standard discount rate. Projects with a positive rate of return are then candidates for funding and construction.

The second investment is in the human organization needed to operate and maintain the constructed system. This process occurs year after year. Here the principles are not as well understood and are almost never taught in schools of engineering or planning. These principles relate to the design of rules that enable organizations to sustain themselves and the physical capital to which they are related over a very long time frame. Frequently, the farmers are expected to do this themselves. And farmers' organizations are frequently successful in overcoming the serious problems of collective action that they face in finding ways to make continued investments in the operation and maintenance of the irrigation works on which they depend. Many irrigation systems, however, do not generate the expected long-term returns due to a failure to achieve an effective organization. The costs of making these investments are almost never taken into account in designing the physical works nor in calculating the overall rate of return for investing in irrigation infrastructure. In this chapter, we study both aspects of the investment in irrigation infrastructure and some of the many challenges that must be overcome to make these investments sustainable, efficient, and equitable over the long term.

5.1 *Evaluative criteria*

Water is a scarce resource. Whether to allocate water in one form or another, in one area or another, and in a large or small quantity, are important questions needing careful consideration. Harvesting activities discussed in the earlier chapters require certain harvesting implements. But those are frequently of nominal cost. In comparison, the cost of physical structures required for harvesting of water resources are substantial, making the question of efficient use of scarce financial resources a serious issue. Whether to use available finance for this or that project is a crucial decision. In short, both water and finance—as scarce resources—should be used efficiently. Some projects have faced severe problems because of waterlogging and salinity. By now, it is well recognized that the meaning of efficiency should be extended to include environmental features as well. Efficiency, therefore, should be understood as optimum use from both economic and ecological perspectives. It should not violate the sustainability of the ecosystem.

Sustainability has an additional meaning here. Investments once made cannot be easily dismantled. Even though considerable care is taken when planning for the construction of new infrastructure facilities, the operation of actual irrigation systems frequently differs substantially from initial plans. This is especially a problem in regard to large-scale, donor-assisted, government-constructed irrigation systems in developing countries. The construction of irrigation infrastructure, once commissioned, does not guarantee proper performance. All irrigation systems need recurring inputs for their operation and maintenance. If operations and maintenance activities are not undertaken properly, the irrigation system itself collapses in a short while. As was noted in Chapter 1, sustainability in general means that "human resource users avoid major disruptions and collapses and can hedge against instabilities and discontinuities."

Three criteria are frequently used for assessing investments in irrigation infrastructure: efficiency, sustainability, and equity. They are complexly related. The models described in this chapter give us an excellent opportunity to explore the interrelations of the three criteria. We have developed a series of simplified models of the stages of development that such projects go through. The first model—the planning model—is based on the biophysical factors that affect the *design* of the physical structure of an irrigation system. The planning model represents the type of thinking that design engineers use when planning and constructing a project. In this Stage I Model, the technological milieu and local physical conditions are considered to be of major importance. We introduce here how decisions about the construction of irrigation systems take into consideration issues of efficiency and sustainability (including both infrastructure and ecosystem meanings of sustainability).

We then examine a Stage II Model, representing the operational phase. When the planned physical infrastructure is implemented, it now functions under real conditions that are not identical with those assumed in the planning stage. Therefore, the actual working of an irrigation system after commissioning may differ from the way it was designed to operate. Water availability is often not the same as planned. The estimate of the average flow, which is the basis of project formulation, might have been an overestimate or an underestimate. It might have changed during the functional stage of the project because of changes in the upper reaches of the source flow. Further, project estimates are normally based on some average flow estimates. Year-to-year variations occur. Crop output is determined not only by water availability, but also by such features as soil quality, nutrient availability, and evapotranspiration effects. As the heat index varies from year to year, evapotranspiration rates vary along with it. The productivity of water does not remain the same every year. Even if water supply is regular and absolutely certain, the consequences may vary for many other reasons.

Probably the most significant difference is that the water is actually used by the farmers. Water users may not take exactly the same actions as envisaged in the planning stage. The differences need not occur only because of their inefficiencies, though that cannot be ruled out. Sometimes, farmers have very different objectives from those of the project planners. Changes from planned operations may happen for a positive reason—the farmers may be trying to cope with altered or varied situations. At other times, the farmers may use their wider knowledge or adopt new techniques. The actual performance of irrigation systems, therefore, differs from the envisaged plans. Sometimes the project may perform better than originally projected, sometimes worse. We examine some of these differences in a Stage II Model.

In discussing the operational phase, we then focus on how individual decisions and collective action are related in this kind of coupled system. No irrigation system operates without continuous effort by individuals to maintain the constructed and natural capital and to distribute the water itself to those who apply it to their fields. Is it not possible for them to extend these efforts to organize themselves to solve their joint problems as well? We include an institutional investment variable so that we can systematically explore the relationship among building more effective institutions, the rules that are developed, and their effects on the performance of the infrastructure investments in terms of both efficiency and sustainability. In addition, we explore the effects of institutional investments on equity among farmers.

5.2 *Growth pattern: crop-water relationships*

As in Chapter 3, we start with the growth rate of the natural product. The harvested product in this chapter is the crop output of irrigated agriculture. Crop productivity depends on a multitude of factors like area cultivated, crop chosen, water used, climatic peculiarities of the region, the soil type, the rainfall or humidity pattern in a particular year, farmer practices like weeding or the use of fertilizer, and even sudden calamities like pest attacks. Many of the factors listed in the "global" section of Chapter 9 affect productivity. Many researchers have studied the effects of one or the other of these factors. But the literature does not contain a single production function that combines all these studies. Therefore, it was left to our discretion to adopt one. In Appendix 5.1 we explain the grounds for adopting a particular expression for the crop-water relations.

To keep the present model simple we have chosen only a few factors. The two major inputs of agriculture, land and water, are obvious choices. With two factors, output can even be expressed in terms of a Cobb-Douglas production function, or something similar. However, such a production function implies that output can be increased indefinitely by increasing either or both of the factors. Farmers everywhere know that this is not the case. Generally, as water application increases, crop yields increase until reaching a maximum output. After that, continued water application results in an eventual decrease of production. We have used a function that depicts this relation. We have disregarded other factors and have explained the behavior in terms of evapotranspiration alone.

Evapotranspiration refers to two physical processes—evaporation and transpiration—that result in the loss of water to the atmosphere. Evaporation entails the loss of water from the soil surfaces, while transpiration refers to the loss of water from the cuticle or stomatal openings on leaves of plants. Because of the difficulty in estimating evaporation and transpiration separately, the two processes are almost always lumped together under the term "evapotranspiration." Accounting for the loss of over two-thirds of the precipitation in the U.S., evapotranspiration is a critical component of the hydrological cycle and an equally important factor in crop production (Dunne and Leopold, 1978).

We express the crop production function (or crop-water relationship) as:

$$Y = Y_m\,[1 - f(1 - (w/AWC)^n)]; \quad 0 \leq Y \leq Y_{m,} \tag{5.1}$$

Where Y is the output per unit of land, w is the water applied per unit of land, Y_m is the maximum of Y corresponding to w = AWC, the maximum amount of water the soil can hold, and f and n—two parameters that are held constant in the Stage 1 Model. The parameter, f, represents the yield response factor based on the type of crop and affects the steepness of the initial response curve as shown in Figure 5.1. We have used the value of 1.5—after consulting empirical data about the yield response of wheat in the central United States. Since the data are not representative of all crops, nor are they available in terms of price, we used a round figure closest to the exact parametric value obtained by fitting the data. The variable, n, depends largely upon meteorological and environmental conditions. This parameter has been set at 0.2 for all of the models discussed herein as it represents the midpoint of known empirical relationships (See Appendix 5.1). As shown in Figure 5.1, the crop water relationship is defined only within a range $0 \leq Y \leq Y_m$ due to the fact that productivity can never be negative or above the possible maximum.

5.3 *Stage I model: the engineering planning stage*

As introduced at the beginning, ecosystem use through irrigation activity is conditioned by the completion of a physical infrastructure. Interaction begins at the construction stage. Using their technical expertise, and the necessary data about the locality, experts formulate technically feasible project proposals. Whether a project is worth undertaking or not must depend on its being rewarding enough, in some sense. In most cases, project selection involves the use of a cost–benefit analysis of financial investments for the infrastructure. If the proposed project is expected to bring in sufficient returns, it is considered economically viable. If it is not sufficiently rewarding, it is rejected and the proposed course of interaction is abandoned. In this section, we introduce the underlying consideration in planning and selection procedures for irrigation projects. In keeping with the framework presented in Chapter 1, the general structure of the Stage I Model is to have three basic subsystems: an ecosystem, a human system, and an interaction system (see Figure 5.2). We have followed the same scheme.

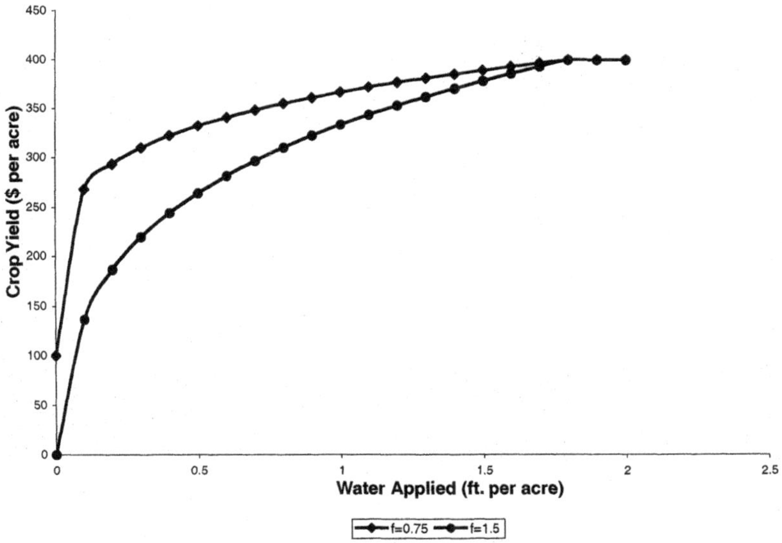

Figure 5.1 As water application per unit of land increases, the yield rate of a crop increases, rapidly at first, then gradually, finally reaching a maximum after which further application of water does not increase yield. Two hypothetical crops, with diffent types of response to water application, are shown. For both of them, the efficient water application rate in this graph is 1.8 ft. per acre producing a maximum yield per acre worth $400.

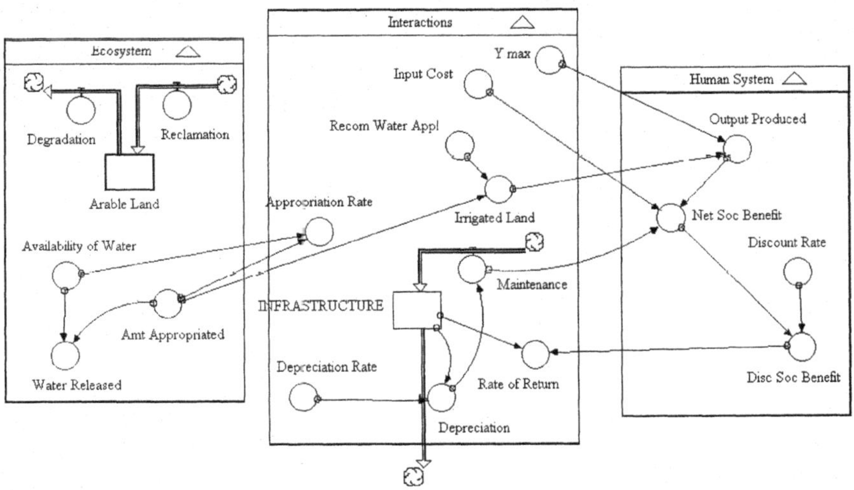

Figure 5.2 The general framework for linking human systems and ecosystems is here applied to the initial design phase of an irrigation system.

5.3.1 Ecosystem

As indicated, we consider only three ecosystem variables: land, water, and evapotranspiration. The model shows how they interact in project proposals. Arable land, not the geographical area per se, is of relevance to irrigated agriculture. Arable land in the locality is a stock; it may be increased by reclamation and may decrease due to soil degradation or erosion. Water availability in a region is a flow variable. Usually, only a part of the total water available in a region is appropriated, and thus the remainder flows downstream.

5.3.2 Interaction

An irrigation project transforms parts of the arable land to irrigated land. Crops grown on this land bring in benefits, while the creation and cultivation of such land involves costs. The benefits from an irrigation project arise not by the mere fact of water appropriation, but by its successful use for various purposes. Usually agricultural and other kinds of benefits (both direct and indirect) are considered. In this simple model, we consider only the direct (agricultural) benefits. The interaction sector therefore includes both irrigation and agricultural activities.

From the considerations of the local conditions and technological possibilities, a proposal envisages the crop that is likely (or desirable) to be grown on the irrigated land. At the planning stage it is assumed that the water appropriated will be used efficiently. That is, the farmers will apply water at the optimum level based on crop type, average evapotranspiration losses, and other environmental conditions (e.g., soil type). Proposals usually specify a recommended level of water application after considering several factors, particularly evapotranspiration rates in the locality. This can be compared with the recommended harvest rule of Chapter 3. Although the water application rate is not the same as the harvest, in the model used for project formulation it determines the crop harvest. The recommended water use per acre (Recom_Water_Appl) is a figure close to, but lower than AWC, the maximum amount of water the soil can hold. The estimate of water appropriable by the project (Amt_Appropriated) divided by the recommended level of water application per unit of land leads to the estimate of arable land that can be irrigated by the project. This is called service area or command area (Irrigated_Land).

$$\text{Irrigated\_Land} = \text{Amt\_Appropriated}/\text{Recom\_Water\_Appl} \tag{5.2}$$

By assuming that water will be used efficiently in operations stage, at the rate of Recom_Water_Appl, project proposals predict productivity, usually close to the possible maximum output per acre (Y_max). From this and the estimates of Irrigated_Land, total agricultural benefits arising from the project are estimated. The cost includes not only the cost of irrigation, but also that of agricultural inputs—seed, fertilizers, etc. To account for these costs, projects include another parameter, Input_Cost (material input per output), usually a constant.

A proposal must not only be technically sound, but should also include estimates of cost of construction (Infrastructure). This is a cost that is incurred only once, but can produce benefits year after year provided the infrastructure is sustained by regular maintenance. Deterioration may also occur due to natural calamities or vandalism. Project proposals normally assume that such incidents will not happen. But depreciation in the natural way is taken into account. Usually it is assumed that regular depreciation occurs at the rate of about 10% of the capital cost. Sustaining the project at its full capacity over the long term implies that exactly this amount is used for maintenance every year. So project proposals adopt a simple assumption that:

$$\text{Maintenance} = \text{Depreciation} \tag{5.3}$$

This simple assumption is rarely met and is one among many assumptions that may turn out to be problematic in the day-to-day operation of an irrigation system over time. We will focus extensively on its implications and consequences in the next section.

Release and distribution of water requires personnel and cost for engaging them. This is called costs for operation of the system. One may regard the variable "Maintenance" in the model as "Operations and Maintenance" (O & M). In most public irrigation projects it is expected that farmers themselves will, at a minimum, meet the O & M costs, for otherwise, the irrigation project becomes an indefinitely long-term responsibility of the government. For the farmers to meet the costs of operating and maintaining the system, receiving contributions from all the beneficiaries is imperative and becomes a collective action issue. At the project formulation stage, however, it is hoped that no free riding will occur and that farmers will voluntarily meet the maintenance costs.

The interactions sector also includes the amount of water appropriated. This is determined by the water availability in the region and the capacity of the physical infrastructure. The capacity is expressed by a constant:

$$\text{Appropriation\_Rate} = \text{Amt\_Appropriated}/\text{Availability\_of\_Water} \tag{5.4}$$

Water appropriated will be less if the physical structure deteriorates over time. Since this is one of the major problems of irrigation systems, we have dealt with this in the Stage II Model. At the proposal stage it is assumed that the physical structure will be sustained at its full potential by regular maintenance. Hence Amt_Appropriated is taken as a constant at this stage.

The final variable is the all important rate of return. This is determined by the discounted social benefit and the amount of infrastructure:

$$\text{Rate\_of\_Return} = \text{Disc\_Soc\_Benefit}/\text{Infrastructure} \tag{5.5}$$

This is the indicator by which the interaction possibility is evaluated. An economically viable project has a rate of return above 1. The larger the rate of return, the more attractive the project is as an investment opportunity.

5.3.3 Human system

The project generates some benefits through the use of water by human agents. This part therefore, is the concern of Human System. Since it is assumed that the water appropriated by the project would be used efficiently, as optimal allocation per unit of land, it follows that estimates of the yield rate are also given by this relation, as Y_max. Thus, the project proposal estimates that:

Output_Produced = Irrigated_Land*Y_max (5.6)

Out of the financial return from output produced (since all output-related equations are denominated in the same financial unit) some amounts have to be paid for material inputs like seeds and fertilizers, and the maintenance costs have to be borne. Thus, a project estimate shows the direct benefits as:

Net_Soc_Benefit = Output_Produced*(1-Input_Cost)-Maintenance (5.7)

The net social benefit occurs year after year against a project. Therefore, its discounted average is considered to be the net social benefit. Usually the bank interest rate is taken to be the discount rate for future benefits. In STELLA notation,

Disc_Soc_Benefit = NPV (Net_Soc_Benefit*Discount_Rate) (5.8)

In the design phase, the estimate of discounted social benefit is compared with the cost as in equation (5). If the rate of return is sufficiently high, then the project is accepted. Otherwise, it is not.

5.4 An overview of major assumptions

A list of the relations and values of parameters used in the Stage I Model can be found in Appendix 5.2. We have considered a hypothetical project costing $300,000 that appropriates 720 acre-feet of water annually from an estimated flow of 2000 acre-feet through the region. The project planners estimate that 400 acres of land can be irrigated efficiently by the project. In other words, the project has been planned to appropriate 36 percent of water flowing through the region, to extend irrigation facilities to 20 percent of arable land in the region, and to generate substantial benefit every year. For this demonstration model we assume that only one crop is produced per year. If the project is sustainable, it produces net benefits year after year, which may be aggregated, after discounting, to obtain total net benefit. With a 5% discount rate,

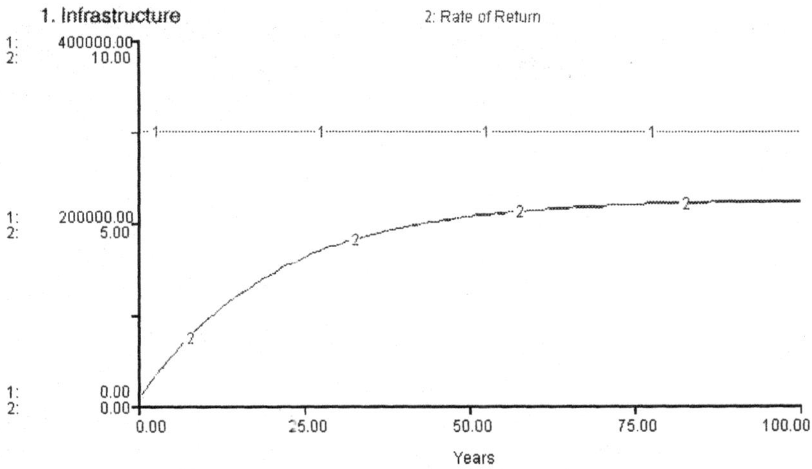

Figure 5.3 Results from a STELLA run with all assumptions consistent with those made in the design phase. The depreciation of the system is assumed to be offset by active maintenance, so that the value of the infrastructure ($300,000) would remain the same over time. The rate of return grows rapidly in the early years and reaches 5.15 after 50 years and then slowly grows to 5.58 after 100 years.

the rate of return of the hypothetical project is 5.15 for a 50-year time horizon and 5.58 for a 100-year time horizon (Figure 5.3).

5.5 Stage II model: the operational stage

Let us assume that the above project was actually constructed exactly as planned. Such is not always the case, but we are not interested in studying differences due to bad construction or poor planning. A project, no matter how well-planned, is based on many approximations and expectations. Realities differ. As a result, actual performances may significantly differ from the projected performances. The Stage II Model (Figure 5.4) depicts the scenario that the project faces after being commissioned. The following is a list of differences from the Stage I Model (see Appendix 5.3 for more details).

- **Water availability:** Project planning was based on some estimate of water flow. Even if the average flow during the operational phase remains the same, the actual flow may vary from year to year. We now model water availability in the region including variations in flow.
- **Water appropriation:** During the planning stage it was assumed that the irrigation infrastructure would always be maintained properly. If the infrastructure is not maintained properly, its capacity to appropriate water from the source would not be the same as before. We include the possibility of the irrigation system losing water appropriation capacity because of deterioration.

- **Output variations due to climatic conditions and crop choice:** An implicit assumption in the planning stage was that a known crop will be grown, under some fixed conditions, producing predictable yield (Y_max). That does not happen in reality, and yield rates vary. In order to predict the yield rates we explicitly include the crop water relations in the Stage II Model. We allow for variability of the parameters f and n in the crop-water relations so that different crops could be grown and climate and environmental conditions could change.
- **Inclusion of individual beneficiaries:** In the operational phase, water appropriated by the irrigation infrastructure is shared by individuals. They are the ones who ultimately determine the performance of the irrigation system. We have therefore, included individual farmers with their objectives and decision-making processes. In this model we have introduced two individuals (or, if desired, teams of individuals), a substantial simplification of the reality of most systems. Farmer A and Farmer B are independent individuals in many regards, but also share some concerns which can be considered to be part of a social system. Thus we split the Human System into three subsystems: the individual farmers (A and B) on the one hand, and a Social System on the other. The Social System represents the institutional relationships among the farmers.
- **Uncertainty in maintenance:** The configuration for the individual sector is the same for all individuals. It should show their possession of resources: land and water. Knowing these, and the crop chosen by them, outputs can be predicted by using the appropriate crop-water relation function. From their outputs, they meet their domestic expenses, costs of agricultural inputs, and maintenance dues for the irrigation structure. A rule must exist that fixes maintenance dues to be paid by each farmer. Whether they pay the dues or not is a matter of their independent decisions—we have included decision functions for payments of maintenance dues in the individual sectors. Maintenance is now a variable composed of the maintenance costs paid by both farmers:

$$\text{Maintenance} = \text{A:\_Actual\_Maintenance} + \text{B:\_Actual\_Maintenance} \tag{5.9}$$

  There is no guarantee that maintenance will be sufficient to affect depreciation, as was envisaged in the planning stage.
- **Institutional investment opportunity:** The Social System includes the distribution parameters, how land and water resources are divided between the two individuals. In addition, we have introduced an institutional investment variable that was not in the project formulation stage. It reflects the amount invested by farmers for development and maintenance of institutional structures to govern processes related to the collective aspects of irrigated agricultural systems. This can be

thought of as the value of the time and energy that the farmers invest in meetings and other occasions to design their rules, to reflect on their performance, and to sanction those who do not follow their rules.

After changing one or more of the parameters and relations of the Stage II Model to reflect the variations likely to be seen in field settings, it is now possible to study how performance differs during the actual operation of systems located in diverse ecological, economic, and social environments.

5.6 *Achieving optimal performance in an operational phase: the benchmark*

If throughout its period of operations the project faces an "ideal" condition, then the predicted rate of return is achieved. What would be necessary in the operational stage to achieve close to optimal performance? The ideal conditions involved assuming that the flow of water, the type of crops grown, and the meteorological conditions remain as specified in the planning documents; the area brought under irrigation is exactly 400 acres divided between the two individuals; each farmer receives water proportionate to the farmer's irrigated land; and both farmers pay their maintenance dues regularly. Meeting these conditions means that four of the variables in the Stage II Model must have the following values since these were implicit in the calculations made by the planners: (1) Water_Flow_ Variation = 0; (2) n = 0.2; (3) f = 1.5; (4) Ins_Inv_per_farmer = 0.

To achieve optimality, some of the other variables can take on any of several values within a defined range. For example, the engineering plans were based on the presumption that the total land to be irrigated was 400 acres. This could be divided between the farmers in multiple ways so long as the total irrigated land did not exceed 400 acres. One way of setting up a benchmark operational model is to make the following assumptions: (5) irrigated land of each farmer (Irrigated_Land) = 200 acres, (6) output produced is the maximum, and each farmer receives a fixed share of water (A:_Fixed_Water_Share) = X, where X is a user defined share. Likewise, the total arable land could have been divided between the farmers in various ways. Here too we could have assumed that it is divided between the two equally. Later, we want to study the consequence of a farmer converting even more of his holdings to irrigated land than in the initial plans. Converting 1,000 acres when the area planned for acreage is only 200 acres seems too extreme. So we have kept Farmer A's holdings at a level which is not very far from 200 acres, as (7) A:_Land_Share = 1/4.

Finally, we make some assumptions about the expenditure patterns of the farmers. The domestic expenses of farmers have a crucial bearing on the issue of maintenance. If a farmer is always in deficit he will often fail to meet his share of the maintenance dues. Besides, a farmer in this model may stand for a group of farmers, each one having domestic requirements. A farmer

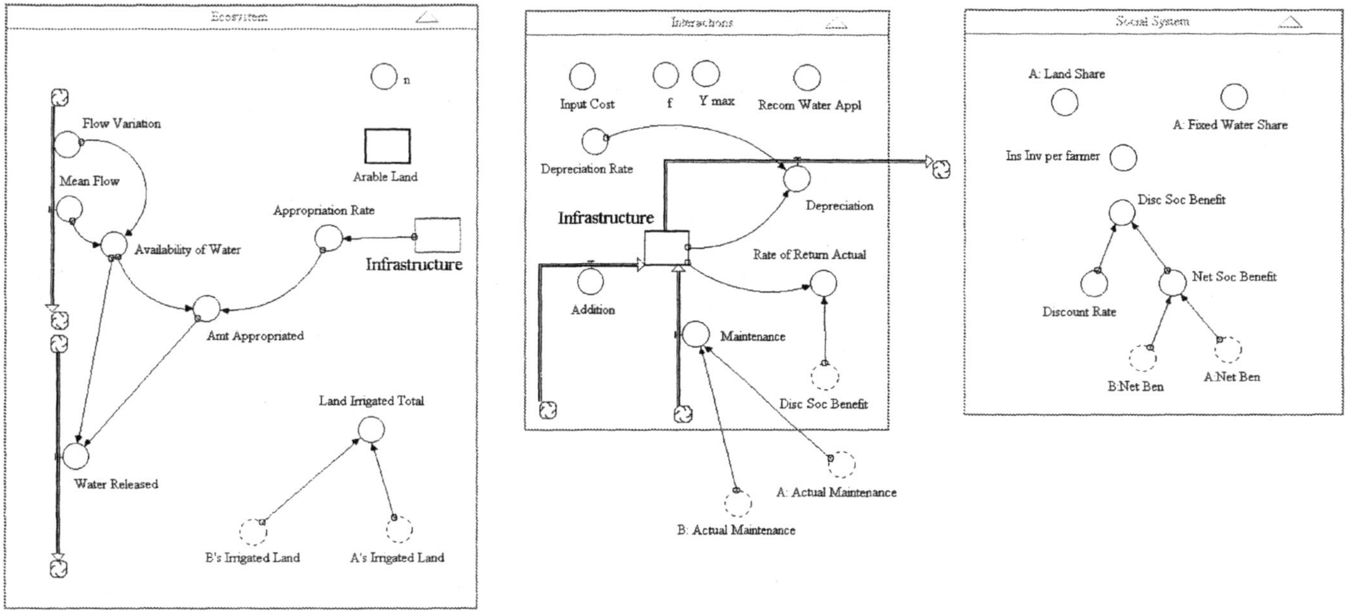

Figure 5.4a The general framework for linking human systems and ecosystems is now applied to the operational stage of running an irrigation system. Here, the revised Ecosystem, Interaction, and Social System are shown.

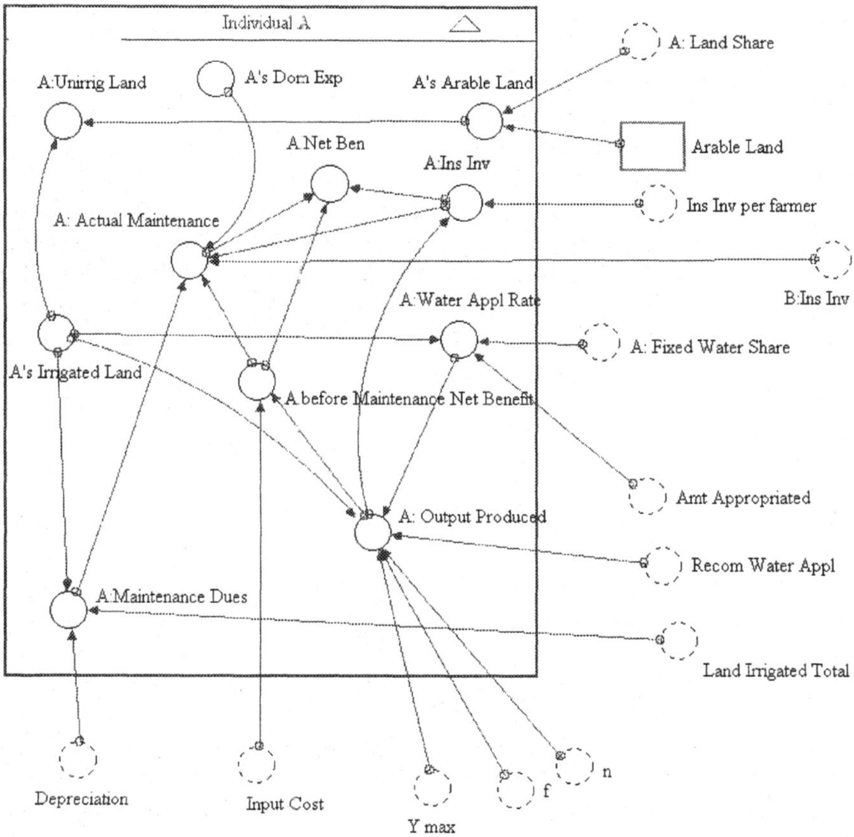

Figure 5.4b The stocks and variables affecting Farmer A's actions are shown. (A parallel subsystem for Farmer B also exists in this model but is not shown since it is identical in structure to that of Farmer A.)

having a large surplus after meeting his domestic needs will not face much of a problem even if output suffers badly in a year. We can make only an arbitrary assumption since expenses of farmers vary widely. We have assumed that the (8) domestic expenses of each (Dom_Exp) farmer = $35,000. With these settings, running the Stage II Model reproduces the output scenario envisaged in the planning stage. The irrigation infrastructure is perfectly sustainable, the total output is the same as that predicted by the Stage I Model, and the rate of return is identical with that estimated at the planning stage. The resulting graph from STELLA runs using these assumptions is shown in Figure 5.3. This model could, of course, have been computed analytically without embedding it in a STELLA model. We do so here as the foundation for the analysis that builds on this Stage I Planning Model.

5.7 Typical problems involved in an operational phase

Rarely will an operational phase so closely approximate initial plans. Many problems are faced by the farmers in trying to keep a system going, and there are also many opportunities that were not originally planned. The Stage II Model can be used to examine a very large number of diverse and complicated problems and opportunities. To illustrate some important questions, we have chosen to examine five possibilities in the remaining sections of this chapter. Two of these are variations in the ecosystem: the variability of the water supply and of the rate of depreciation due to storms and other external shocks. The other two reflect the variety of actions that the farmers take: the possibilities of increasing the land to be irrigated and the decision whether to pay maintenance dues or not. The fifth problem relates the head-tail syndrome of irrigation systems, whereby one farmer is located higher in the system and obtains his water before the second farmer.

5.7.1 Water availability varies

The calculations made during the planning stage were based on an average value for the expected water flow into the irrigation system. Even if this amount is available in the aggregate the actual pattern of distribution may not be supportive of best use. Crops require water at certain stages of growth. If the supply is not available at that period, productivity is affected. Also, water is frequently spread unevenly over the command area. Actual supply can fall short of average expected supply. Performance variation of irrigation systems due to such factors has become a major concern in recent years (Small and Svendse, 1992; Sengupta and Sampath, 1995). Our simple model does not show within-year variations, but we can use this model to study the effects of inadequate water availability and inequity in distribution over the command area. Here we discuss the former.

We will consider the year-to-year variations that are to be expected in most environments, in spite of the average remaining the same. We now assume that the Water_Flow_Mean is as predicted in the planning stage (2000 acre-feet), but it varies within a range from 1/3 to 2/3 percent (Flow_Variation = 0.33; = 0.50; = 0.67). As we see in Figure 5.5, the output of Farmer A (and thus also of Farmer B since they are initially symmetric with one another) now fluctuates as would be expected. Also, output approaches zero as the variance in water availability increases. The dramatic fall in agricultural output is not just due, however, to the variance in water supply by itself. In those years with low agricultural output, farmers are much less willing to contribute to the maintenance of the irrigation system and to its long-term viability. If farmers only contribute to the maintenance of the irrigation system after they have met their normal domestic expenses—the assumption that has been made so far—variations in water availability make a major difference in whether the irrigation system is maintained or not, and thus in the long-term output achieved by the farmers.

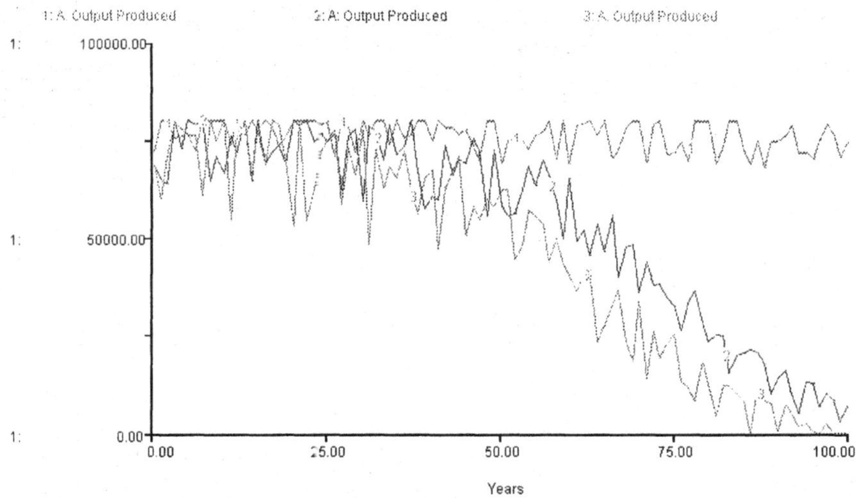

Figure 5.5 Effect of natural variations of water on the output of Farmer A. Water flow varies by 33% (line 1), 50% (line 2) and 67% (line 3). For small variations agricultural output fluctuates but does not decrease. For large variations it decreases substantially due to the decay of the irrigation infrastructure resulting from occasional failures of maintenance.

In Table 5.1, we present the average value of irrigation infrastructure for each decade after construction depending on how much water supply varies while holding all of the other assumptions of the benchmark Stage II Model constant. If the water supply varies by only one-third, the infrastructure remains in relatively good shape over a 100-year period based on farmers contributing to maintenance once their domestic expenses are met. If, on the other hand, water availability varies by 50%, the value of the infrastructure decreases by almost two-thirds after 70 years of operation, and to only $23,780.23 by the end of a century. And, if water availability varies by 67%, the value of the infrastructure is reduced by approximately one-half after only 30 years, and is down to one-fourth of its value by 50 years.

5.7.2 Depreciation higher

In many irrigation systems, considerable damage occurs occasionally (e.g., systems in hilly areas). A disaster of great magnitude is something that may overwhelm farmers' capacities to cope, but smaller events may be tackled by them. In other words, the average depreciation rate is higher than 10% and calamities occur from time to time that exceed the average.

To bring the possibilities of such calamities into the model, let us consider that the magnitude of the depreciation varies between 10 to 25% and that such calamities occur approximately once in every 5 years. Otherwise, steady depreciation at the rate of 10 percent occurs every year

Table 5.1 Value of Irrigation Infrastructure Depending on Variation in Water Availability

(Based on an average of 10 runs for each level of variation)

Variation in Predicted Flow

Year	0.33	0.50	0.67
0	300,000.00	300,000.00	300,000.00
10	292,733.96	278,938.49	261,627.32
20	288,611.11	260,342.80	215,022.65
30	283,333.06	235,266.97	169,927.47
40	277,346.07	206,329.30	127,483.27
50	271,733.07	181,108.91	75,115.50
60	265,103.76	153,023.36	37,278.47
70	258,914.37	117,813.90	18,010.23
80	255,122.61	78,757.67	6,504.95
90	249,552.93	45,715.52	2,268.14
100	244,159.83	23,780.23	790.85

as before. This is represented in STELLA by replacing the old equation Depreciation_Rate = 0.1 by:

Depreciation_Rate = 0.1*(1 + PULSE(RANDOM(0,1.5), 5, RANDOM(0,10))) (5.10)

The first term in the BUILTIN function PULSE is for magnitude, the second specifies when the first pulse occurs, and the third, the waiting time. None of these is definite. We have treated magnitude and regularity as random along with the occurrence of the first pulse at the fifth year. Figure 5.6 shows the fluctuations of depreciation rate and its consequence on the irrigation infrastructure. As can be seen, the value of the infrastructure steadily erodes as farmers are not likely to produce a higher agricultural product in the same years as the calamity. Thus, they do not make all of the repairs needed to make the system operate at full capacity again. As minor calamity follows minor calamity, the system itself slowly erodes.

5.7.3 Farmers' decision: how much to irrigate

Another typical event that occurs after an irrigation system has been constructed is that the farmers want to allocate water to more land than originally planned. Let us assume that Farmer A owns 500 acres of land. In an operational phase, he has the option of using the same amount of water allocated to him to irrigate the whole or any part of his land. In the benchmark, we assumed he used the water on 200 acres of land. In Table 5.2, we present the annual productivity per acre and the value of his annual output of his applying the *same* quantity of water to differing amounts of land ranging from 100 acres up to 500 acres (without fluctuations in water availability).

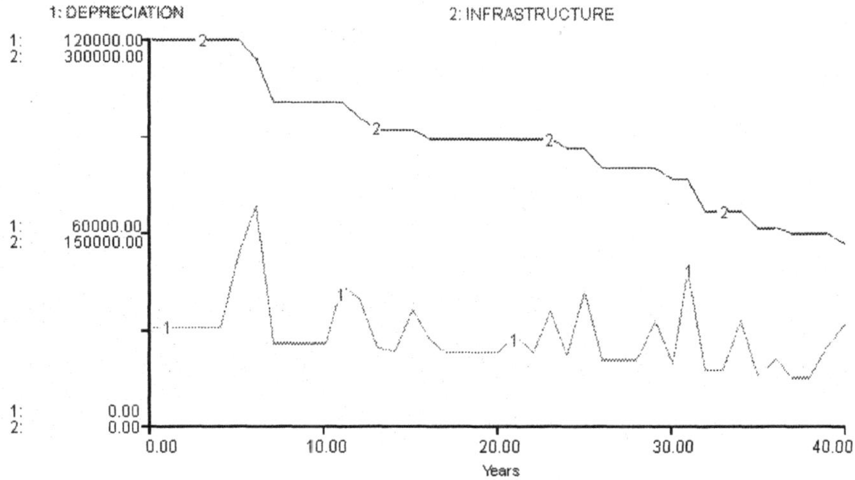

Figure 5.6 Effect of variation due to natural calamities: Natural calamities, modeled here as occurring randomly once every five years, damage the infrastructure beyond the capabilities of farmers to keep it fully repaired after every calamity.

Although the productivity of land decreases by a thin spreading of the available water over wider areas, Farmer A's total product increases. Thus, if he acts rationally, a farmer who owns more land is not likely to follow the recommended water application rate. He benefits more by converting as much of his arable land as possible to irrigated land. In this instance, the actual rate of return for the entire project increases because of a wise decision made by a farmer. Not all unexpected events after the planning of an irrigation system are detrimental to its overall productivity.

5.7.4 The threat of free riding

If the farmers are narrowly rational and myopic, they will be tempted to free ride on the maintenance of an irrigation system. Since this linked human-ecosystem is an ongoing enterprise that could continue for a long time into the future, most farmers would not regularly shirk on their maintenance responsibilities in those years where their agricultural productivity was sufficient to meet their domestic expenses. Most would realize that if they refused to contribute anything to the cost of maintenance, others would soon do the same, and the irrigation system would not last very long at all.

Table 5.2 Productivity and Output for Farmer A Depending On Number of Acres Irrigated

Land irrigated by A (acres)	100.00	200.00	300.00	400.00	500.00
Annual Output ($)	40,000	80,000	105,979	128,932	149,766
Productivity ($ per acre)	400.00	400.00	353.26	322.33	299.53

However, a strategic farmer might be tempted to be somewhat erratic in paying for maintenance. Let us think of this as one of the farmers paying less than the required maintenance dues in an unpredictable fashion given by:

A:_Actual_Maintenance = RANDOM(0, A:Maintenance_Dues) (5.11)

Table 5.3 shows the results in regard to the net annual benefits for both farmers when Farmer A does not pay his full maintenance dues regularly. If he were to pay these expenses regularly as in the benchmark model, he would have an annual return of $38,333 every year. If Farmer B were to continue to pay his maintenance dues regularly, Farmer A's irregular payments do not immediately have a major impact on the condition of the irrigation system. The system slowly deteriorates. Even after the 10th year, with Farmer B paying dues regularly, Farmer A's net benefit is $45,033, which is substantially above $38,333.

We have also shown the net benefits received by Farmer B. His maintenance dues are the same and he continues to pay them. Farmer B's net benefit declines slightly because of deterioration of the infrastructure. If Farmer B adopted the same strategy as Farmer A, the consequence is shown in Table 5.4.

Since the maintenance cost paid varies randomly in our model, the net benefits received by each farmer fluctuate considerably. Yet, almost in all years except the 10th year for Farmer B the net benefit for both farmers is above $38,333, the amount they would have received with a regular payment of maintenance dues. Hence erratic payment of the maintenance cost is a "rational" choice for both farmers for the first 10 years after an irrigation system is constructed. In the process, however, the infrastructure depreciates due to inadequate maintenance. Table 5.4 is one of the many possible patterns of payment of dues. Since actual maintenance is paid at random, the pattern will vary from run to run. Because of the range of the random variable, the expected payment of maintenance cost is exactly one-half of the maintenance

Table 5.3 Effect of One Farmer Paying Maintenance Dues Erratically

Year	Maintenance Dues	Actual Maintenance of Farmer A	Net Benefits for Farmer A	Net Benefits for Farmer B
0	15,000	13,161	40,172	38,333
1	14,908	2,019	51,217	38,327
2	14,264	7,679	44,853	38,268
3	13,934	2,994	49,169	38,229
4	13,387	3,028	48,506	38,147
5	12,869	9,172	41,747	38,050
6	12,685	9,388	41,307	38,011
7	12,520	7,359	43,134	37,973
8	12,262	7,677	42,496	37,911
9	12,032	6,142	43,741	37,850
10	11,738	4,471	45,033	37,766

Table 5.4 Both Farmers Are Erratic in Paying their Maintenance Dues

Year	A: Net Benefits	B: Net Benefits	Depreciation	Maintenance	Condition of Infrastructure
0	44,358	42,991	30,000	19,318	300,000
1	40,500	40,059	28,932	24,951	289,318
2	46,735	44,291	28,534	14,045	285,337
3	40,986	43,502	27,085	18,940	270,849
4	47,426	41,256	26,270	13,793	262,704
5	39,533	45,900	25,023	15,532	250,226
6	44,377	46,652	24,074	8,748	240,736
7	45,926	44,596	22,541	7,253	225,411
8	46,692	42,403	21,012	6,573	210,123
9	46,367	44,811	19,568	2,384	195,683
10	40,141	38,260	17,850	13,578	178,499

dues, which is 10% of the value of the infrastructure. Thus, on average, one-half of the infrastructure will have undergone decay after ten years. A more severe effect will be felt in later years. Therefore, once a longer period is considered the picture appears very different. Figure 5.7 shows the same situation over a 100-year period. After the fluctuations in net benefits for both farmers for the first few years, everything goes into a steady decline. As can be seen in the figure, after 25 years, both farmers regularly receive less than $38,333, the amount they would have made if they had both made regular payments. Further, the value of the infrastructure has declined to about one-fourth of its initial value by the 25th year. Note, however, that even with this drastic reduction in the net benefits achieved by the farmers in the years following the 10th year, as well as the major decline in the infrastructure, the rate of actual return remains very close to the benchmark model—a subject that we will address below.

5.8 *Investment in irrigation institutions*

One of the ways that farmers can deal with the problem of inadequate maintenance is by investing in institutions that insure that all farmers contribute their share of maintenance costs. In field settings where effective rules are designed and followed to a substantial degree, farmers make a recurring set of investments in an institution. These investments may take several forms. Farmers attend regular meetings where modifications of the rules they use regarding allocation of water as well as requirements for maintenance are proposed, discussed, and decided upon. When one farmer breaks a rule others must decide whether and how much to sanction this farmer. Farmers must also spend time determining how to undertake the maintenance of their jointly used irrigation system. Each of these activities, and many others, requires different types of rules and institutional arrangements. In developing the Stage II Model, we included a variable for institutional investments (Ins_Inv_per_farmer) to represent this capacity to invest in institutions that

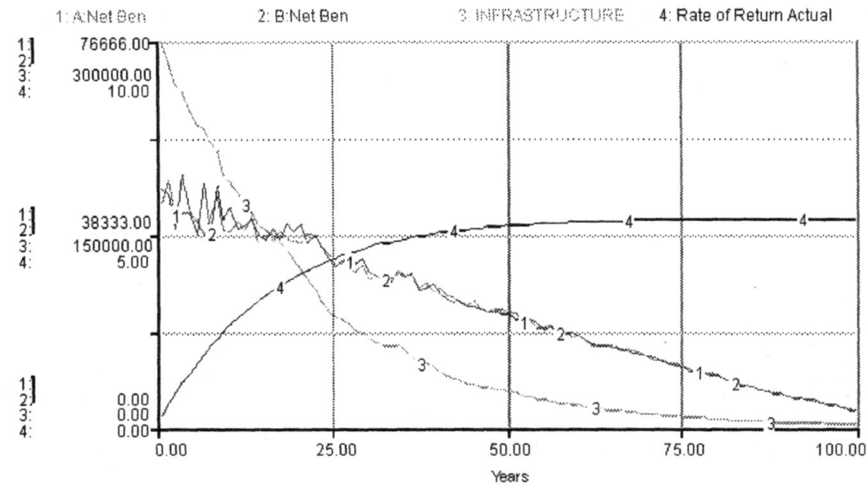

Figure 5.7 Effect of human strategic behavior: If neither of the farmers pay maintenance dues regularly, their net benefits (lines 1 and 2) are higher than planned in the first 10 years before declining dramatically. The value of the infrastructure (line 3) deteriorates substantially almost immediately. Even though the infrastructure suffers about 75% decay in just 30 years, and net benefits are reduced appreciably, the rate of return (line 4) is not affected substantially (compared with the rate of return shown in Figure 5.3).

keep the maintenance of the system performed. We have so far assumed it to be zero. Now we will examine how such investments help to solve the problems discussed above.

It is hard to determine how much it costs to create an effective farmer organization that helps farmers to overcome the temptation not to contribute their full share of the maintenance of their joint system. If farmers share most values and are very similar in terms of their assets, the costs may be less than when there is considerable asymmetry among them. To examine how these investments may affect system performance we compare results for when no investment is made, when the investment is moderate (or 20% of the total maintenance costs), and when the investment is very high (or 40% of the total maintenance costs). The costs of establishing and sustaining an institution is shared between the two, such as the total maintenance cost. Thus, we will study three levels of investment: 0, $3,000 and $6,000 a year. The result of an institutional process is the implementation of and compliance with a set of rules that compel both to pay their share of maintenance cost. A strict rule (e.g., that in all circumstances farmers have to pay the maintenance cost) will establish the effectiveness of institutional rules even more than what we have achieved here. But such a formulation is too mechanical. To be more realistic, we have allowed for some leeway—if the farmer's product is excessively low in one year, or always, then he does not invest in institution activities, which are costly and will further reduce his benefits in that year of distress. In turn, when farmers do not invest in their institution, the rule to compel them to pay their maintenance dues is not effective during that period. This is expressed as:

A:Ins_Inv = IF(A:_Output_Produced/10 < Ins_Inv_per_farmer)
THEN(0) ELSE(Ins_Inv_per_farmer) (5.12)

An equivalent expression exists for Farmer B.

Let us now recall how decisions are represented about farmers' decisions to pay maintenance dues. Equation 11 described the case where Farmer A behaves in an irregular manner, but in the basic model we have described actual maintenance as:

A:_Actual_Maintenance = MAX(0,(MIN((A:before_Maintenance_Net_
Benefit − A's_Dom_Exp),A:Maintenance_Dues)) (5.13)

The relatively complicated formulation serves to restrict the values within the possible range: 0 and Maintenance_Dues. The relation says that within this range, a farmer pays only as much as he can after meeting his other costs, including his domestic expenses. To include the possibility of making an institutional investment, we have modified these actual maintenance expressions as:

A:_Actual_Maintenance = IF(A:Ins_Inv = 0 OR B:Ins_Inv = 0)
THEN (MAX(0,(MIN((A:before_Maintenance_Net_Benefit-A's_
Dom_Exp),A:Maintenance_Dues)))
ELSE(A:Maintenance_Dues) and (5.14)

A:_Actual_Maintenance = IF(A:Ins_Inv = 0 OR B:Ins_Inv = 0)
THEN(RANDOM(0, A:Maintenance_Dues))
ELSE(A:Maintenance_Dues) (5.15)

This form has been used uniformly beginning with the benchmark model. In the benchmark model, however, we assumed Ins_Inv_per_farmer to be 0. If the institutional investment made by both the farmers is positive, however, both of them do pay the required maintenance dues. Otherwise, deficit or irregular payments may occur. Institutions are cooperative processes; they are not built by individuals alone. Hence we have included within the IF clause, a form that means an institutional investment has no effect if only one of them participates in the institutional process and the other fails to do so.

Suitable institutional investments can ameliorate these instances of decreasing optimal performance. Since all the cases occur because of the decay of the infrastructure, an effective rule that leads all farmers to invest in maintenance will help to check the reduction in performance. We will show only one case. We have selected a complex case for this purpose where both ecological and human problems occur. The flow variation is assumed to be

0.20, a positive but not excessive level so that this is not a particularly problematic system from an ecological point of view. In addition, we assume that the farmers do not completely refuse to pay maintenance, but that both pay erratically as discussed above. The first two columns of Table 5.5, where the institutional investment is assumed to be zero, demonstrate that these two assumptions which are relatively realistic in field settings may easily bring disaster to a system. The next two pairs of columns show the situation after varying levels of institutional investment. In both cases the infrastructure is made sustainable over time. The net social benefit fluctuates slightly, but remains relatively high even after 100 years. In this instance, a lower level of institutional investment—at the equivalent of 20% of the maintenance costs—insures long-term sustainability of the system and of the net income for the farmers. A higher level—at the equivalent of 40% of the maintenance costs—also insures the sustainability of the physical system, but at a reduced level of net social benefit. Under investing in institutional arrangements is a disaster. Over investing is costly.

5.8.1 Sustainability and efficiency

In the above sections we considered the impact of institutional investments on sustainability, but not the effect of these investments on the rate of return. If we consider the problem of coping with weather and other calamities by increasing the level of institutional investment, we determined above that irrigation systems that were otherwise unsustainable are made sustainable with an investment in institutions that insure that maintenance is performed. Such investments are costly and have an adverse effect on the economic rate of return as normally computed, even though without the investment the infrastructure would deteriorate and farmers' incomes would approach zero. Table 5.6 shows the rates of return for the three alternatives discussed above and shown in Table 5.5.

Even though the irrigation system survives for a much longer period of time, the additional cost incurred every year—especially in the early years—for institutional activities adversely affects the rate of return. A similar pattern occurs when investments are made in institutions to assure sustainability in order to cope with any of the other kinds of problems discussed above. We have, of course, assumed quite high levels of institutional investment. It would be a trivial problem, however, if we did not consider a realistic cost for all of the time and effort that is involved in creating effective rules, monitoring performance, and imposing sanctions. Given the adverse impact on how economic efficiency of an investment in infrastructure is normally computed—the overall rate of return—one must question whether it is meaningful to invest in institutions. Such investments may ensure sustainability and higher levels of individual farmer returns, but these benefits are not reflected in how the rate of return is computed.

Actually, whether institutional investments are useful or not is a question that is highly sensitive to the perspective one takes about the future in terms

Table 5.5 Value of Infrastructure and Net Social Benefit as Affected by Three Levels of Institutional Investment

Year	Inst. Inv. = 0		Inst. Inv = $3000		Inst. Inv. = $6000	
	Value of Infrastructure	Net Social Benefit	Value of Infrastructure	Net Social Benefit	Value of Infrastructure	Net Social Benefit
0	300,000	90,374.42	300,000	70,666.67	300,000	64,666.67
10	166,121	75,244.19	300,000	70,666.67	300,000	63,327.13
20	97,100	69,759.03	300,000	70,666.67	300,000	64,416.44
30	54,995	54,819.30	300,000	69,436.03	300,000	64,666.67
40	31,272	45,562.17	300,000	70,666.67	300,000	64,666.67
50	19,374	41,918.27	300,000	70,666.67	300,000	61,288.26
60	13,108	28,827.59	300,000	65,733.03	300,000	64,666.67
70	7,767	24,511.00	300,000	70,666.67	300,000	58,577.21
80	3,982	15,233.42	300,000	69,313.31	300,000	64,666.67
90	2,345	7,314.29	300,000	70,666.67	300,000	64,666.67
100	1,478	−46.35	300,000	70,666.67	300,000	61,140.39

Table 5.6 The Rate of Return with Different Institutional Investments

	Zero	$3,000	$6,000
After 50 Years	5.10	4.69	4.26
After 100 Years	5.24	5.07	4.62

of (1) the discount rate and (2) the time horizon for aggregation. In Table 5.5, net benefits decrease as the infrastructure deteriorates. The net present value of future benefits depends, however, on the time perspective taken and the weight given to the future stream of benefits. A high discount rate does not give much weight to what happens in the distant future. But a low discount rate means that the future is considered more important. With a discount rate of 5.0%, the rate of return does not increase perceptibly even if the project has a much longer life. But with lower discount rates, a longer life has much greater significance. In fact, as is shown in Figure 5.8, when the discount rate is set at 2.5%, a reversal of the order of outcomes occurs before 100 years. In this case, it is more efficient to make the investment in institutions than to avoid these costs entirely. So if the time horizon is a 100 years or beyond, and the discount rate is very low, sustaining the infrastructure through recurring institutional investments is also economically efficient, as shown by the higher rates of return. In fact, it may be highly rewarding. For a discount rate of 1% the rate of return will increase steadily even after a 100 years (not shown). If one takes a longer horizon, the rate of return for sustaining the infrastructure with institutional investment may be several times more than the rate of return for not incurring the higher institutional investments. In all, we find that analyses of the sustainability of an infrastructure and its capacity to generate income for the farmers is in conflict with the traditional ways of measuring economic efficiency for higher discounting of the future. Investing in institutions is a very efficient strategy for achieving high economic rates of returns whenever one attaches importance to the future, in mathematical parlance, when one uses a low discount rate and a long time horizon.

5.8.2 Equity, sustainability, and efficiency: the head-tail syndrome

Last, we focus on a problem faced in every irrigation system. A farmer whose land is located at the head reach of tributaries can easily withdraw disproportionately higher shares of water than authorized, depriving the land located at the tail reaches of the system. This head-tail syndrome is one of the most common problems of irrigation management. Obviously, excess water appropriated by the head reach farmers affects equity in water distribution, but the adverse effect is not restricted to the question of equity. We will show that it may affect both sustainability and efficiency.

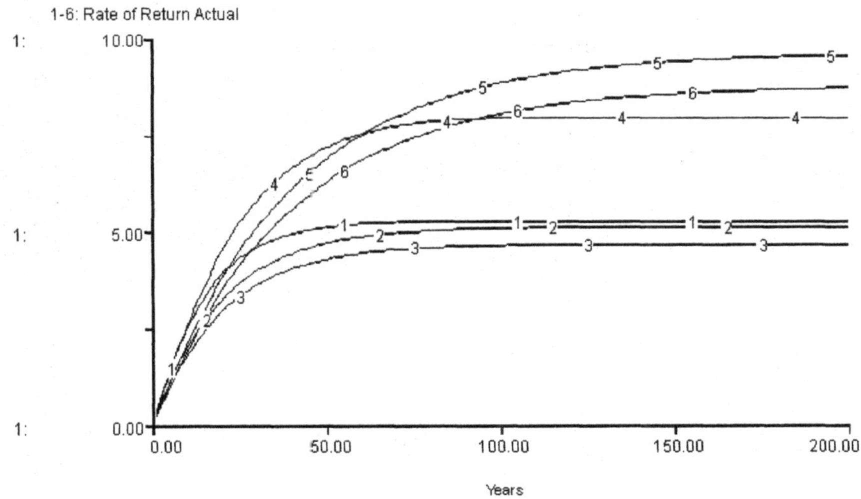

Figure 5.8 Significance of discount rate and time horizon: Line 1 shows the rate of return when neither of the farmers pay maintenance dues regularly (same as line 4 in Figure 5.7). Lines 2 and 3 show the rates of return for institutional investment by each farmer at \$3,000 and \$6,000 respectively. Although the irrigation infrastructure is sustainable (not shown in figure) because of institutional activities, the rate of return decreases due to the increased costs of institutional investment. The rates of return are recalculated with a lower discount rate (2.5%) than the original 5%. Lines 4, 5 and 6 correspond to lines 1, 2 and 3 respectively when the discount rate is 5%. Note the reversal of order in the third decade. When the future is discounted at a lower rate, institutional investments leading to the sustainability of infrastructure and higher net returns for the farmers has a much higher rate of return than with the higher discount rate.

In our model, each farmer was entitled to receive a fixed share of water. The share for Farmer A was A:_Fixed_Water_Share and for Farmer B, (1−A:_Fixed_Water_Share). Multiplied by the Amt_Appropriated, depending on the condition of the infrastructure, these determined the total water available to each farmer to irrigate his Irrigated_Land. Thus, water use intensity of each farmer, which determined the productivity of his crop output, was given by:

A:Water_Appl_Rate = Amt_Appropriated*A:_Fixed_Water_Share/
A's_Irrigated_Land (16)

B:Water_Appl_Rate = Amt_Appropriated*(1-A:_Fixed_Water_Share)/
B's_Irrigated_Land (17)

In the benchmark model we had assumed that the farmers take just their due shares. To explore the head-tail syndrome, we change some assumptions. We have designated Farmer A as the farmer having land at the head of the tributary who may appropriate more water. We introduce a new variable,

A:_Water_Overused, and assume that A appropriates 50% more water than his equitable share. Farmer B, having land at the tail end, receives only whatever is released by A. Hence, the water use intensities are now given by:

A:Water_Appl_Rate = Amt_Appropriated*(A:_Fixed_Water_Share*
(1 + A:_Water_Overused))/A's_Irrigated_Land (18)

B:Water_Appl_Rate = Amt_Appropriated*(1-A:_Fixed_Water_Share*
(1 + A:_Water_Overused))/B's_Irrigated_Land (19)

The consequence of this new set of relationships is illustrated in Figure 5.9. In only a short thirty years, the adverse effect of allowing the head-end farmer to grab all the water he needs can be seen. For the first 15 years, Farmer A's net benefit continues to rise while Farmer B's net income is much lower and does not vary. Thus, the inequitable results are the most prominent feature of the dynamics for the first 15 years. Although, by assumption, Farmer A and B had identical endowments and in most of the previous models had benefited and suffered in identical manners, in this case we find Farmer A increases his net income substantially to the detriment of Farmer B, the tail-end farmer. But this is only a short run result. Farmer B receives less water, produces less, and finds it difficult to pay his share of maintenance dues. As maintenance is not accomplished, the infrastructure deteriorates and finally reduces even Farmer A's benefits. Thus, inequitable distribution of water ultimately affects the very sustainability of the project. And from the foregoing discussions, this, in turn, reduces economic efficiency in a perspective where future is considered sufficiently important. Thus, when the future is considered important, the three criteria of efficiency, sustainability, and equity are closely related. Improving one improves the other as well.

In this case, as well as those discussed above, introducing effective institutional rules leads to the sustainability of the project. A key question, however, is what kind of institutional rule would work where Farmer A is substantially advantaged by his physical location. Surely every contingency does not call for the same rule. In the previous sections, the problems arose because of non-maintenance, and the institutional rule was one of compelling all members to meet the maintenance expenses. Different kinds of rules are used to meet different kinds of problems.

By using an institutional rule that compels both the farmers to pay the maintenance dues regularly, the inequity in distribution of water and in net benefits will persist, but the infrastructure will not deteriorate. But the same old rule, which had an effect only if both farmers invested in it, cannot be used. We cannot assume that Farmer A will adopt it, for he could have achieved its results without such an investment simply by refraining from excess water appropriation. There are several alternative institutional arrangements for this problem, such as monitoring and restrictions, rotational irrigation, etc. Farmer-enforced monitoring and restrictions have fre-

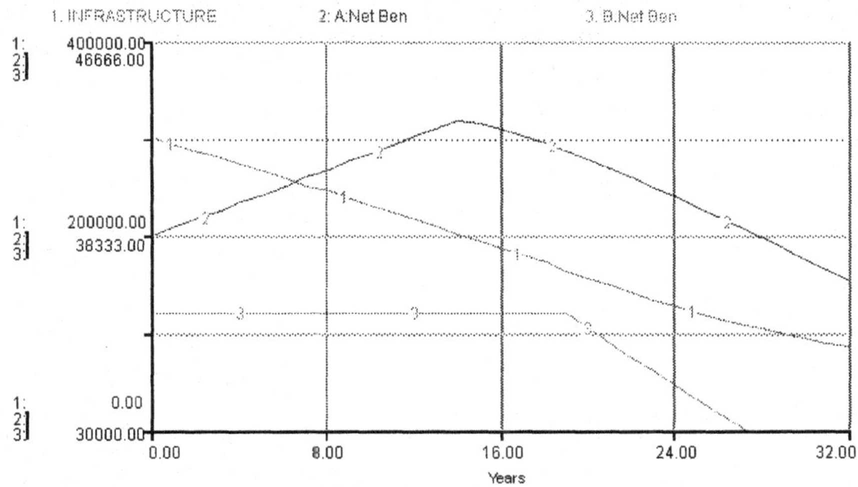

Figure 5.9 The farmer at the head reach of the irrigation system (Farmer A) takes 50% more water than his due. In a decade and a half, Farmer A's net benefit increases to the detriment of the tail-end farmer. The infrastructure steadily deteriorates. In less than 20 years both farmers are worse off than they would have been if Farmer A had taken his authorized quantity of water.

quently been successfully used in overcoming this syndrome (Sengupta, 1991; Tang, 1992; Lam, 1998). Observers have sometimes puzzled over some of these rules as observed in the field since the costs of enforcing them are frequently borne primarily by those who already are adversely affected by the inequitable distribution of water.

An alternative institutional rule consists essentially of Farmer B monitoring and restraining Farmer A from an overuse of water. Farmer B invests his time and effort for this purpose, agreeing that A has no interest in doing this. This situation requires substantive modification in the Stage II Model. We discuss these in Appendix 5.4. Farmer A may even exercise some levels of self-restraint, if there is an institutional arrangement whereby the farmers regularly discuss mutual problems. This may be easier to appreciate if we understand A and B as groups (or arrays) of farmers at the head and tail end of a system rather than a single farmer. Farmers may not only discuss alternative use patterns, but they may also introduce imaginative rules such as rotational irrigation. Even if these rules only succeed partially, there may be substantial gains in equity, sustainability, and efficiency.

In Figures 5.10–5.12, we assume that Farmer B alone invests in the institution, and thus achieves a partial success. Farmer A does not invest anything, but is partly persuaded to oblige. The paths marked 1, 2, 3, 4, 5 in all three figures correspond to 50, 40, 30, 20, or 10% overuse of water by Farmer A. Farmer A still over uses water in all of the cases examined, but the adverse effect on the infrastructure decreases steadily with the success achieved, though not eliminated completely. Figures 5.11 and 5.12 show how much net benefits respond. Here too we have selected the range for ease of comparison. Beginning with a slight increase in the net benefit of the privileged party, in the very early years most cases of water overuse by the head reach farmer results in ultimate decrease of net benefit for both farmers. But this adverse effect can be deferred by increasing the success of monitoring and restraining strategies. In fact, Farmer B can succeed dramatically (Figure 5.12) in postponing his downfall by attaining even partial success in restraining the head reach farmer. There are varieties of institutional rules to cope with varieties of problems. One may use this model to study others that are not discussed here.

5.9 Conclusion

In this chapter, we have used the framework presented in Chapter 1 to explore several central questions related to the linkage of human systems with complex constructed physical systems that are themselves closely linked with natural capital. We have illustrated the process of designing complex and costly engineering tasks such as the construction of irrigation infrastructure. At the time of design, future human decisions are presumed to follow a fixed average pattern that will generate optimal returns given the assumptions built into the engineering designs. Not only is the physical world modeled as a determinate system, the humans who will use the future irrigation system are presumed to plant only the crops that were used in the

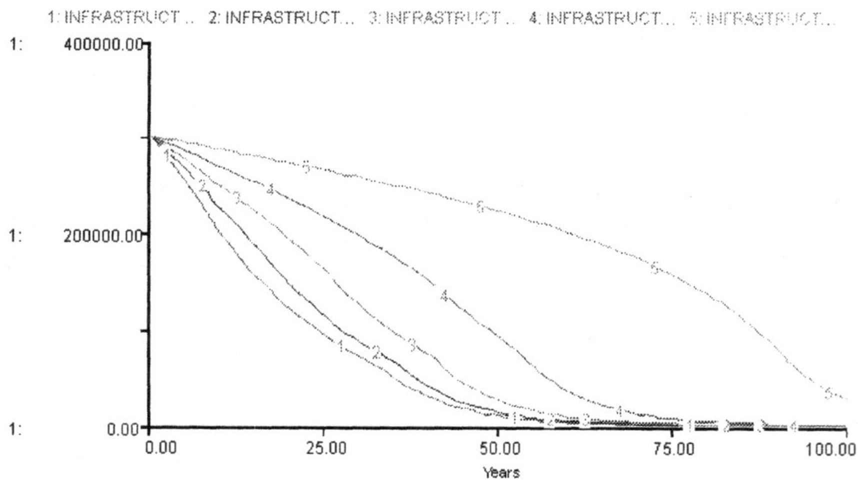

Figure 5.10 Institutional investment of $3000 by the tail end farmer (Farmer B) alone partially restrains the head reach farmer. Infrastructure deteriorates, but not as fast (line 1, excess water appropriation equals 50%). Lines 2, 3, 4 and 5 show lower rates of excess water appropriation, at 40%, 30%, 20% and 10%, respectively.

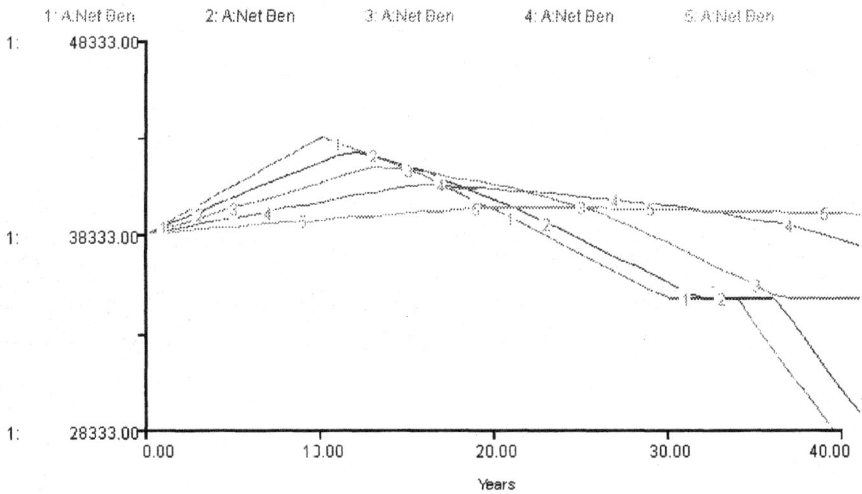

Figure 5.11 Net benefit situation faced by the head reach farmer (A) when partially restrained from excess appropriation of water. Lines 1 to 5 correspond to excess appropriation rates of 50, 40, 30, 20, and 10%, respectively.

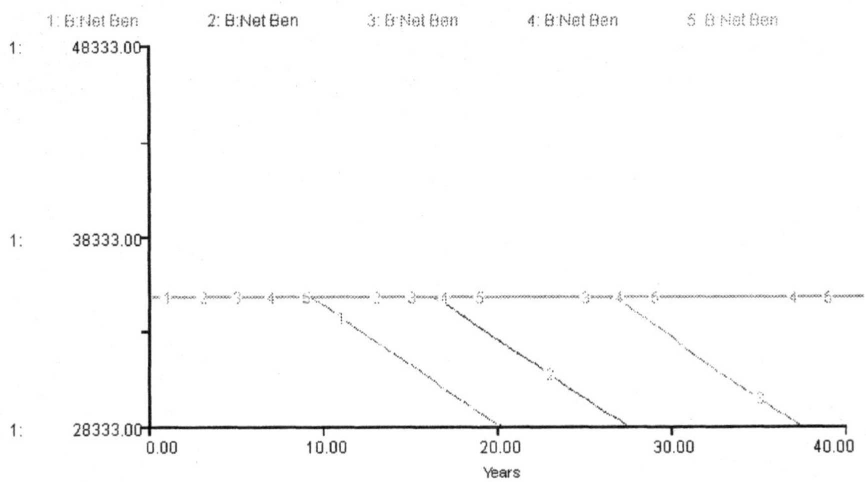

Figure 5.12 Net benefit situation faced by the tail end farmer (B) when he succeeds partially in restraining the head reach farmer from excess appropriation of water. Lines 1 to 5 correspond to excess appropriation rates of the head reach farmer at 50, 40, 30, 20, and 10%, respectively. Note, Farmer B can maintain his net benefits at a steady level for longer and longer periods by restraining the head reach farmer through institutional activities.

design of the project, to irrigate only the pre-determined command area of the design, and to face a world in which the average parameters occur each and every year. In this world of fixed parameters—including the preferences and strategies of the humans involved—the formula for calculated expected economic efficiency of a major investment in physical infrastructures is well established.

Reality, however, is never so fixed. Whether or not the farmers actually build a particular irrigation system or not, they are frequently left with the problem of how to operate and maintain it over many years. In doing so, they are faced with a wide diversity of uncertain events including variations in water supply, in factors affecting the deterioration of the system (such as severe storms), in the amount of land irrigated, in the level of free riding, and in the amount of water taken by farmers located advantageously. The farmers must cope with problems of sustainability, efficiency, and equity in their efforts to find ways of maintaining a system over time. At times, farmers find better ways of managing their own land that have no externalities and lead to a higher net return for themselves, and thus, for the system as a whole. But many of the possibilities that occur after the design and construction of a system increase the risk that farmers do not maintain the irrigation system and that the system itself deteriorates over time. Farmers can, however, make investments in building more effective local institutions that increase the probability of sustaining the infrastructure and the net income of the farmers themselves over time. We have shown, however, the somewhat paradoxical result that using currently accepted discount rates leads to a conclusion that the economic rate of return is higher when farmers do not make such an investment and allow a system to deteriorate. This is because the normal discount rates do not give much weight to the long-term sustainability of a system over time and its capacity to generate income for the farmers over many years. The higher costs in the initial decade or two are given much more weight in the calculation of an economic rate of return than are the many decades in which farmers could be making high net benefits if they had been investing in institutions that resulted in the sustainability of the system.

Thus, these models have enabled us to explore some important questions of how the evaluative criteria of sustainability, efficiency, and equity are related when considering the long-term operation of humanly constructed infrastructures that affect the way a bio-physical ecosystem operates. Results that generate inequitable outcomes may lead to system deterioration because those who are disadvantaged do not continue to maintain the system or invest in institutions that would solve the maintenance problems. In some cases, however, those who are disadvantaged may go ahead and bear a larger proportion of the institutional costs in order to generate a more equitable distribution of benefits and costs. Under some circumstances, the disadvantaged are better off for taking this initiative.

One of the most interesting conclusions to come from our analysis is that major decisions about investing in infrastructures and the institutions to make them work must be based on estimated costs and benefits that weight the future more than is frequently done in normal benefit-cost analysis. To make such systems work over time requires that those operating these systems take a long-run perspective and make investments that make economic sense when the impact of these decisions on system performance in the decades ahead are included in the analysis.

5.10 Appendix 5.1

5.10.1 Crop water relations

Crop output depends primarily on the crop type, area cultivated, and water used. These are, however, only the basic inputs. Productivity is affected by other factors as well, including physical/biological conditions such as climate, soil type, and even pests. It is also affected by farmer practices such as weeding or the use of fertilizer. To keep the present model simple we have chosen only three inputs: land, water, and evapotranspiration. For convenience, the crop-water relation has been expressed as output per unit of land or productivity (denoted by Y) by combining two of the inputs, land and water, as water applied per unit of land or intensity of water application (denoted by w). It is a common practice to reduce production functions, including the Cobb-Douglas production function, to constant returns to scale with respect to one of the factors.

Accounting for actual evapotranspiration effects on crop yield is complex. For example, not only do evaporation rates depend on varying climate conditions, but transpiration rates even differ for varying growth stages of the same crop (Hillel, 1987). Overall, crop yields are expected to increase with corresponding increases in water application until reaching a maximum evapotranspiration rate. Prior to reaching ideal water conditions, evapotranspiration rates largely represent transpiration—an integral process in plant growth that is positively correlated with high crop yields. This ideal maximum evapotranspiration rate is often equivalent to the potential evapotranspiration rate (PET), the amount of water lost when plants are not limited by water supply. As expected, prior to reaching this maximum rate, plants are not obtaining full water needs and thus produce lower yields. A simplified formula proposed by Doorenbos and Kassam (1979) to evaluate crop yields relates an increase in yield as actual evapotranspiration (AET) increases:

$$Y/Y_m = 1 - [f(1 - (AET/PET))] \tag{5.1.1}$$

This equation thus forms a key foundation for our Stage I Model. As actual evapotranspiration reaches potential evapotranspiration, the proposed relationship shows a linear increase in the proportion of the actual yield (Y) to the maximum yield (Y_m). The f represents the yield response factor based on crop type. However, in reality, studies show that the relationship between crop production and evapotranspiration is not fully linear. After reaching and passing the maximum rate, the relationship between crop yield and evapotranspiration is complicated by the evaporation of water from the soil surface. While transpiration is construed as a plant activity that contributes to yield, evaporation contributes to water loss from the system. This excessive water loss through evaporation does not enhance crop production. In extreme examples, excessive waterlogging may result in increases in carbon dioxide and oxygen deficit that may impair the ability of plants to uptake water and actually lower crop yields (FAO/UNESCO, 1973). The impact of excessive waterlogging will vary with crop type.

In order to better represent the effect of evapotranspiration on crop yields in the model, the AET/PET relationship must be better defined. Despite the complexity of the relationship and limited quantitative information, it is commonly understood that soils must have very low amounts of available soil moisture before there are large decreases in actual evapotranspiration (Dunne and Leopold, 1978). The relationship implies an exponential factor. We assume a simple function explains this behaviour:

$$AET/PET = (w/AWC)^n \text{ where AWC and n are parameters.} \qquad (5.1.2)$$

The constant term AWC refers to the available water capacity, that is, the maximum amount of water the soil can hold considering gravitational effects and soil type. A value of 0.2 is used for the exponent n. It most closely resembles the midpoint of a range of possibilities for the relationship as demonstrated by other research (Dunne and Leopold, 1978).

The combination of equations 2.1 and 2.2 results in a more sensitive crop production equation:

$$Y = Y_m [1 - f\{1 - (w/AWC)^n\}] \qquad (5.1.3)$$

The parameter, f, may lie anywhere between 0 and AWC. For $w = AWC$ one gets $Y = Y_m$. In addition, this mathematical function shows that Y can be increased indefinitely by increasing w. This does not happen in reality, however. Therefore we assume that Y stabilizes at the maximum Y_m once w increases to AWC. Also, there is a minimum. For very small values of w while $f > 1$, the equation gives a negative value of Y. This only shows that for those crops with $f > 1$, a minimum amount of water is necessary for producing any output. Thus we write:

$$Y = Y_m [1 - f\{1 - (w/AWC)^n\}]\ 0 \leq Y \leq Y_z \qquad \text{(Eq. 5.1)} \qquad (5.1.4)$$

Within the permissible range, this production function indicates a declining rate of growth of yields with increasing water application until reaching a maximum crop yield (Figure 5.1).

5.11 Appendix 5.2

5.11.1 Key assumptions used in Model I

The following relations were used in the model:

Arable_Land(t) = Arable_Land(t − dt) + (Reclamation − Degradation) * dt		(2.1)
INFRASTRUCTURE(t) = INFRASTRUCTURE(t − dt) + (Maintenance − Depreciation) * dt		(2.2)
Water_Released = Availability_of_Water-Amt_Appropriated		(2.3)
Irrigated_Land = Amt_Appropriated/Recom_Water_Appl	(eq 5.2)	(2.4)
Appropriation_Rate = Amt_Appropriated/Availability_of_Water	(eq 5.4)	(2.5)
Maintenance = Depreciation	(eq 5.3)	(2.6)
Depreciation = INFRASTRUCTURE*Depreciation_Rate		(2.7)
Output_Produced = Irrigated_Land*Y_max	(eq 5.6)	(2.8)
Net_Soc_Benefit = Output_Produced*(1-Input_Cost)-Maintenance	(eq 5.7)	(2.9)
Disc_Soc_Benefit = NPV(Net_Soc_Benefit,Discount_Rate)	(eq 5.8)	(2.10)
Rate_of_Return = Disc_Soc_Benefit/Infrastructure	(eq 5.5)	(2.11)

The following parameter values were used for the hypothetical project:

INIT Arable_Land = 2000 acres	(2.12)
INIT INFRASTRUCTURE = $ 300000	(2.13)
Availability_of_Water = 2000 acre-feet	(2.14)
Amt_Appropriated = 720 acre-feet	(2.15)
Recom_Water_Appl = 1.8 feet	(2.16)
Y_max = $ 400 per acre	(2.17)
Depreciation_Rate = 1/10	(2.18)
Discount_Rate = 0.05	(2.19)
Input_Cost = 1/3	(2.20)
Reclamation = 0	(2.21)
Degradation = 0	(2.22)

Since this model was established to create the baseline that would be used in estimating the benefit-cost ratio for a project like this, it is not a very "interesting" dynamic model. Equation 2.6 makes Equation 2.2 not really a dynamic equation. In later stages we drop 2.1, 2.2, and 2.6 which make Model II and its variations fully dynamic, showing interesting interactions among variables.

5.12 Appendix 5.3

5.12.1 Key assumptions used in Model II

The model for the planning stage includes several approximations and anticipations. The model for the operational stage provides for possible variations in reality. In the following sections we have introduced the changes that were made in the Stage I Model to arrive at the Stage II Model.

5.12.1.1 The Ecosystem

We have omitted here the question of reclamation and degradation of arable land. There are three major changes:

(a) Water availability:

Water availability in a region is not fixed. We assume that it varies randomly above and below the mean water flow (Mean_Flow) considered during the planning stage, the range being given by a parameter Flow_Variation:

Availability_of_Water = RANDOM((1−Flow_Variation), (1 + Flow_Variation))*Mean_Flow (5.3.1)

(b) Water appropriation:

The Infrastructure was originally planned to appropriate 36% of the water flowing through the region. However, this capacity will not be retained if the infrastructure is not maintained in conditions as good as during the commissioning. In an operational phase, there may occur some loss of capacity due to deterioration. We assume that the irrigation system loses capacity in direct proportion to the deterioration.

Appropriation_Rate = .36*Infrastructure/INIT(Infrastructure) (5.3.2)

The actual water appropriable by the infrastructure at any point of time is then given by:

Amt_Appropriated = Availability_of_Water*Appropriation_Rate (5.3.3)

Hence the water released from the region is given by:

Water_Released = Availability_of_Water-Amt_Appropriated (5.3.4)
(same as 2.3 in Model 1)

(c) Inclusion of variations in meterological conditions:

Agronomic studies of crop-water relations show output variations due to meteorological conditions such as air temperature, sun availability, wind velocity, and overall heat index, all of which affect the amount of water that can be absorbed into the atmosphere. All of these factors play a large role in determining the potential and actual evapotranspiration rates, as well as the relationship between the two. Additionally, the size and frequency of rain events impacts evapotranspiration because it controls water availability. The parameter n in equation (1) may be considered a variable, accounting for this variety of conditions.

5.12.1.2 *Interactions*

(a) Inclusion of crop choice possibility:

The variable f in the crop-water relations equation (1) accounts for type of response to water application obtained from different crops (Figure 5.1). To include the possibility of alternative crops, we include it as a variable in Model II. Two of the parameters included in equation (1) are included in different sectors. Since n accounts for ecological features, it belongs to the Ecosystem sector, while f pertains to crop choice and is thus an interaction decision made by farmers.

(b) Uncertainty in maintenance:

Actual maintenance is no longer assumed to be equal to the rate of depreciation. Maintenance is now a variable composed of contributions made by both farmers:

Maintenance = A:_Actual_Maintenance +
B:_Actual_Maintenance (5.3.5)

(c) Natural calamities:

In one variation of the Stage II Model, Equation 3.18 was changed to equation 10.

5.12.1.3 *Human system*

(d) Inclusion of individual beneficiaries:

The human system now has two components: individual farmers and the social system. The Output Produced and Net Social Benefits are determined at the individual levels. The Social System finds the aggregate of the individual net social benefits as the total net social benefit, on the basis of which the discounted benefits are calculated as before, with the same discount rate. To begin with, we have introduced only two individuals. Many more individuals can be linked in the same way. Moreover, each unit may be redefined to represent a group of farmers.

5.12.1.4 *Social system*

(e) Distribution pattern:

The Social System includes the distribution parameters that divide the land and water resources between the two individuals. These are: (i) A:_Land_Share and (ii) A:_Fixed_Water_Share. Since these are expressed in shares, Farmer B has only the rest of the share.

(f) Institutional investment opportunity:
Institutional investment as a variable may include many things, from costs of meeting, that of monitoring, or indirect costs of rule compliance regarding lost opportunities. The Ins_Inv_per_farmer variable indicates the investment required, which is not necessarily paid. We have used three different levels: 0, 20, and 40% of maintenance cost, shared by both the individuals.

Individual farmers

(g) Land distribution:
Instead of assigning each farmer a set amount of land, we assigned Farmer A a share of the land and Farmer B the remainder. The relative shares of the two farmers can be considered as an equity index.

A's_Arable_Land = Arable_Land*A:_Land_Share (5.3.6)

One has to put some value to the irrigated land. It can be any value between 0 and the total arable land of the farmer. The project proposal envisages that the total irrigated land of the beneficiaries will be equal to what can be irrigated efficiently by the project; 400 acres for the hypothetical project demonstrated here. In reality it may be very different. Once a system is in operation, there is no certainty that the individuals will convert their holdings to irrigated land in the same manner that the project planners had envisaged. In the Stage II Model, the irrigated land of A and B are variables (A's_Irrigated_Land and B's_Irrigated_Land). The rest of their holdings are therefore unirrigated.

A's_Unirrig_Land = A's_Arable_Land-A's_Irrigated_Land (5.3.7)

Farmer B receives the remaining land, i.e., 1-A's share.

(h) The distribution of water:
The other endowment is water. We need to ask how the total appropriated water is shared between the two. When applied on their respective irrigated land, the amount determined the water use intensity. The project formulation is based on an implicit ideal situation where farmers receive fixed shares of water. Hence:

A:Water_Appl_Rate = Amt_Appropriated*A:_Fixed_Water_Share/ (5.3.8)
A's_Irrigated_Land (Eq. 16)

B:Water_Appl_Rat = Amt_Appropriated*(1-A:_Fixed_Water_Share)/ (5.3.9)
B's_Irrigated_Land (Eq. 17)

But one of the farmers may succeed in obtaining more water than his due share by being located at the head reach of the distribution system or simply by threat and power. The appropriate equations must be changed (Eqs, 5.18 and 5.19) in a case of variation in Stage II.

(i) Output produced:

This relation can differ drastically from that of the Stage I Model. There it was assumed that the project will function efficiently and output produced will be equal to Y_max for the recommended crop. In the operational phase the farmers may do better or worse, may choose different crops, and overall efficiencies may be higher or lower. Also, natural conditions differ from year to year. Therefore, in this model we replace the idealized production estimate by the actual production function that responds to varying conditions. Since the production function is specified in terms of crop–water relations, the water application rate is the relevant variable.

A:_Output_Produced = IF(A:Water_Appl_Rate > Recom_Water_Appl) (5.3.10)
THEN(Y_max*A's_Irrigated_Land)
ELSE(MAX(0, (Y_max*A's_Irrigated_Land*(1-f*(1-(A:Water_Appl_Rate/Recom_Water_Appl) ^{n})))))

Equation 5.3.10 is the production function (Equation 5.1) rewritten in the language of STELLA. When the water application rate is high enough to fill the soil's maximum water-holding capacity (as represented by Recom_Water_Appl), crop output is maximized to Y_max. When application rates are less than the soil's maximum water-holding capacity, output will depend on crop response to water application (f), relative soil moisture conditions (Water_Appl_Rate/Recom_Water_Appl), and other outside factors including temperature and humidity (as represented by n).

(j) Rule about maintenance dues:

The total maintenance cost needs to be at least equal to the depreciation of the system to keep it working over time. We divided it between the two beneficiaries. We have to assume some manner of legitimate cost sharing. We assume:

A:Maintenance_Dues = Depreciation*A's_Irrigated_Land/Land_Irrigated_Total (5.3.11)

A similar relation describes the dues payable by B so that together their maintenance contributions equal the depreciation. An alternative assumption would have been that each contribution in terms of a share of water, or that maintenance is divided equally among farmer households. All these rules are found in specific field settings and could be used in the model.

(k) Investment for institutional work:

A farmer is not willing to pay if the output is too low.

A:Ins_Inv = IF(A:_Output_Produced/10 < Ins_Inv_per_farmer) (5.3.12)
THEN(0) ELSE(Ins_Inv_per_farmer)

(l) Maintenance cost paid:
We assume that, if after meeting the material input costs of agriculture and the domestic needs (A's_Dom_Exp and B's_Dom_Exp) the individual finds that he has enough to pay his dues, he pays the maintenance dues. Deduction of the input cost leads to another variable.

A:before_Maintenance_Net_Benefit = A:_Output_Produced* (1-Input_Cost) (5.3.13)

The actual maintenance is described by these three variables as

A:_Actual_Maintenance = IF(A:Ins_Inv = 0 OR B:Ins_Inv = 0) THEN(MAX(0,(MIN((A:before_Maintenance_Net_Benefit-A's_Dom_Exp),A:Maintenance_Dues))) ELSE (A:Maintenance_Dues) (Eq 5.14) (5.3.14)

A variation of this in the Stage II Model is expressed by Equation 5.15.

(m) Net benefits at individual levels:
Knowing how much is produced and the level of the other expenses allows us to calculate net benefits. The form is the same as in the Stage I Model; domestic expenses are not deducted to obtain the net benefits. The form is

A:Net_Ben = A:before_Maintenance_Net_Benefit-A:_Actual_Maintenance- A:Ins_Inv (5.3.15)

The following parameter values of the variables introduced in Stage II lead to an operational situation envisaged by the planners of the hypothetical project:

Flow_Variation = 0 (5.3.16)

n = .2 (5.3.17)

f = 1.5 (5.3.18)

A:_Land_Share: can be any value within a range. (we assumed 1/4) (5.3.19)

A:_Fixed_Water_Share = _ (5.3.20)

A's_Dom_Exp (and also of B) = $35000 (5.3.21)

A's_Irrigated_Land (and also of B) = 200 acres (5.3.22)

Ins_Inv_per_farmer = 0 (5.3.23)

Appendix 5.4

Modification in Model II for head–tail syndrome

We drop two variables: Ins_Inv_per_farmer and A:Ins_Investment by B alone remains, but is now given by a slightly modified equation allowing for no institutional investment when the output is very low.

B:Ins_Inv = IF(B:_Output_Produced < 10000) THEN(0) ELSE
(3000 or 6000) (5.4.1)

A new variable A:_Water_Overused has been included. It is included in the Social Sector alongside the water share of A. As the effect of B making institutional investment increases we assume the value of water overused reduces from 0.5 to 0.1. All the five different levels have been simulated.

We have already discussed in the text that the water application rates vary because of the unequal distribution. These are given as:

A:Water_Appl_Rate = Amt_Appropriated*A:_Fixed_Water_Share*
(1 + A:_Water_Overused)/A's_Irrigated_Land (5.4.2)

B:Water_Appl_Rate = Amt_Appropriated*(1-A:_Fixed_Water_Share*
(1 + A:_Water_Overused))/B's_Irrigated_Land (5.4.3)

Institutional investments no longer directly affect the actual maintenance. Hence:

A:_Actual_Maintenance = (MAX(0,(MIN((A:before_Maintenance_
Net_Benefit-A's_Dom_Exp),A:Maintenance_Dues)))) (5.4.4)

Also since Farmer A does not pay anything for institutional investment his net benefit is given by

A:Net_Ben = A:before_Maintenance_Net_Benefit-A:_Actual_
Maintenance (5.4.5)

Farmer B has to consider his institutional investment cost in both cases. Thus for him the equations are

B:_Actual_Maintenance = (MAX(0,MIN((B:before_Maintenance_
Net_Benefit-B's_Dom_Exp-B:Ins_Inv),B:Maintenance_Dues))) (5.4.6)

B:Net_Ben = B:before_Maintenance_Net_Benefit-B:_Actual_
Maintenance-B:Ins_Inv (5.4.7)

Appendix 5.5

List of variables

Variable	Description
A:_Land_Share	share of Farmer A in total arable land in the region
A:_Fixed_Water_Share	share of Farmer A in the total amount of water appropriated by the irrigation infrastructure
Actual_Maintenance	cost of actual work maintenance
Addition	addition to irrigation infrastructure
Amt_Appropriated	amount (volume) of water appropriated by the infrastructure depending on its existing condition
Appropriation_Rate	proportion of the available volume of water appropriated by the infrastructure depending on its existing condition
Arable_Land	arable land in the region
Availability_of_Water	volume of water available from the source
Before_Maintenance_Net_Benefit	gross benefit net of input cost
Degradation	degradation of arable land
Depreciation	depreciation of irrigation infrastructure
Depreciation_Rate	rate of depreciation of irrigation infrastructure
Disc_Soc_Benefit	discounted aggregate social benefit
Discount_Rate	discount rate for time
Dom_Exp	domestic expenditure of farmers
f	a parameter for crop-output relation reflecting the type of crop chosen
Flow_Variation	annual variation determining water availability from source
Infrastructure	irrigation infrastructure (valued in a monetary unit)
INIT Arable_Land	initial arable land
INIT Infrastructure	initial irrigation infrastructure
Input_Cost	cost of seed, fertilizer, per unit of output
Ins_Inv	institutional investment made by a farmer
Ins_Inv_per_farmer	required institutional investment for effecting a rule
Irrigated_Land	irrigated land owned by a farmer
Land_Irrigated_Total	total irrigated land in the region
Maintenance	total maintenance cost
Maintenance_Dues	the maintenance cost demanded from each farmer
Mean_Flow	annual average water availability in the source
n	a parameter in crop-output relation related to ecological features (rainfall, etc.)
Net_Ben	net benefit of a farmer
Net_Soc_Benefit	aggregated net benefit
Output_Produced	crop produced
Rate_of_Return	rate of return envisaged in the proposal stage
Rate_of_Return_Actual	actual rate of return
Reclamation	reclamation of arable land
Recom_Water_Appl	recommended water application (ft per acre)
Unirrig_Land	unirrigated land of a farmer
Water Released	volume of water released from the region after appropriation by the irrigation infrastructure
Water_Appl_Rate	water application rate (ft per acre) of a farmer
Y_max	maximum output per acre

chapter six

An introduction to mathematical models in fisheries ecology

Carl Simon

Contents

1-5667-0389-1/01/$$0.00+$.50

6.1 Introduction

In this chapter we present an overview of the dynamic models used to study renewable resources. We start with the simplest model (exponential growth), add limits to growth (logistic growth), and then bring in harvesting in various economic settings. Focusing on fishery models, we next expand our focus to multiple species, multiple locations and age-structures. We will use standard mathematical notation while minimizing the mathematical background required. Since most of the modeling in this book uses STELLA, we will describe the STELLA formulation for the dynamic models we discuss. On the other hand, whenever possible we will describe—and sometimes derive—analytic or geometric solutions of the mathematical models we present.

6.2 Exponential growth

We write x_t or $x(t)$ for the size of a population x at time t. Population models focus on the **percent growth rate:**

$$\frac{\Delta x}{x} = \frac{x_{t+1} - x_t}{x_t}$$

The simplest assumption is *constant* percent growth rate:

$$\frac{x_{t+1} - x_1}{x_1} = r \quad \text{or} \quad x_{t+1} = (1 + r)x_t. \tag{6.1}$$

If the time period under study is not one year but a more general Δt, then (1) becomes:

$$\frac{x(t + \Delta t) - x(t)}{x(t)} = r \cdot \Delta t \quad \text{or} \quad \frac{\Delta x}{\Delta t} = r \cdot x. \tag{6.2}$$

For Δt very small, we work with Equation 6.2 as the *differential* equation

$$\frac{dx}{dt} = rx. \tag{6.3}$$

The simplest analogy for Equation 6.1 through 6.3 is the growth of a deposit in a savings account with annual interest rate r. Then, Equation 6.1 is the description of simple annual interest; Equation 6.2 is the description of compounding every Δt of a year; and Equation 6.3 is the description of continuous compounding. The solution of Equation 6.1 is $x_t = (1 + r)^t x_0$; the solution of Equation 6.3 is $x_t = e^{rt} x_0$, where x_0 is the population (deposit) size at time $t = 0$.

The STELLA compartmental formulation of these constant percent growth rate models is presented in Figure 6.1.
The STELLA equation editor would write the system as

$$x(t) = x(t - \Delta t) + r * x(t) * \Delta t. \tag{6.4}$$

6.3 *Logistic growth*

The solutions presented above entail unbounded, exponential growth. This may be a reasonable scenario for a short-term study, but such unlimited growth is untenable in the long run. The most obvious way to change the model is to assume that the percent growth rate decreases as the population increases—say, as a result of decreasing availability of the resources needed to sustain growth. The simplest such formulation is a linearly decreasing growth function x {??} $r - bx$. Then, Equations 6.1 and 6.2 become

$$\frac{x(t + \Delta t) - x(t)}{x(t)} = (r - b \cdot x(t)) \cdot \Delta t, \tag{6.5}$$

and Equation 6.3 becomes

$$\frac{dx}{dt} = x \cdot (r - bx). \tag{6.6}$$

Equation 6.5 or 6.6 is usually called the **logistic equation;** in the fisheries literature, it is called the **Schaefer equation.** Figure 6.2 presents the graph of

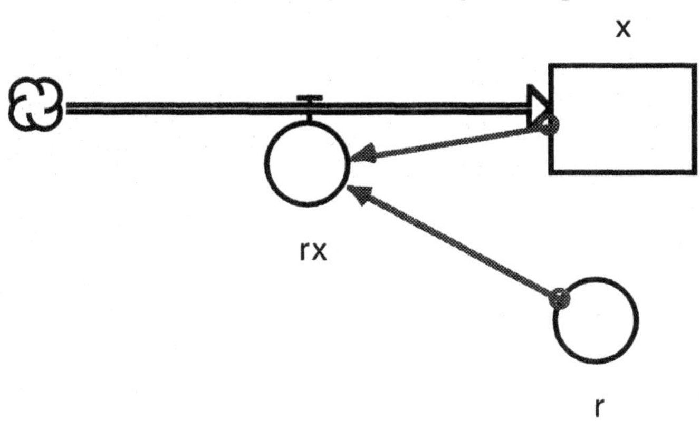

Figure 6.1 STELLA model of exponential growth.

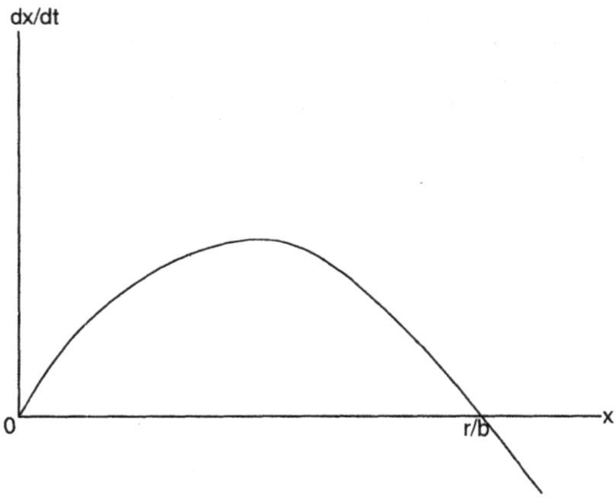

Figure 6.2 Phase line analysis of logistic Equation 6.6.

dx/dt or $\Delta x/\Delta t$ versus x: the right-hand side of Equation 6.6 is a parabola with x-intercept at 0 and r/b. When $0 < x < r/b$, $dx/dt > 0$ and $x(t)$ is increasing in t; when $x > r/b$, $dx/dt < 0$ and $x(t)$ is decreasing in t.

Figure 6.3 summarizes this information. It is called the **phase portrait** of the system and is the geometric "solution" of the dynamic equation.

As the phase portrait indicates, all solutions of the logistic equation (that start with $x(0) > 0$) tend to the equilibrium r/b. In fact, in elementary differential equation classes (Simon and Blume, 1994), one calculates the solution of (6) to be:

$$x(t) = \frac{r}{b}\Bigg/\left[\left(\frac{r}{bx_0} - 1\right)e^{-rt} + 1\right]. \tag{6.7}$$

Since $e^{-rt} \to 0$ as $t \to \infty$, $x(t) \to r/b$ as $t \to \infty$ in Equation 6.7, for any $x_0 > 0$. In mathematical models, this population level r/b to which the logistic system always converges is called the **carrying capacity** and written as K. If one substitutes r/K for b in (5,6), one obtains the more common form of the logistic equation:

$$\frac{dx}{dt} = rx\left(1 - \frac{x}{K}\right) \tag{6.8}$$

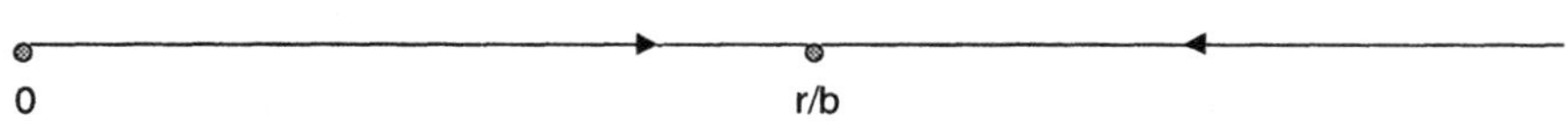

Figure 6.3 Phase diagram of logistic Equation 6.6.

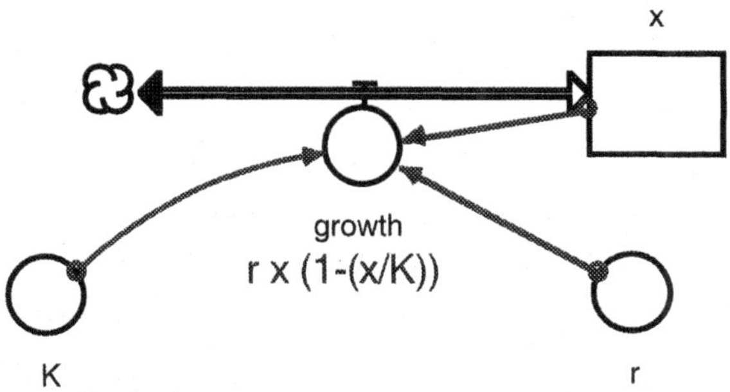

Figure 6.4 STELLA model of logistic Equation 6.6.

The STELLA equation editor would write Equation 6.8 as

$$x(\mathrm{t}) = x(t - \Delta t) + r * x(t) * \left[1 - \left(\frac{x(t)}{K}\right)\right],$$

corresponding to the compartmental model in Figure 6.4.

6.4 *Periodic and chaotic populations*

Technically, the analytic solution Equation 6.7 and corresponding geometric solution in Figure 6.3 apply only to the (continuous-time) differential equation of the logistic expression Equation 6.6. The discrete time logistic equation (with $\Delta t = 1$), which we rewrite as:

$$x_{t+1} = x_t + rx_t\left(1 - \frac{x_t}{K}\right), \tag{6.9}$$

behaves similarly only when $0 < r < 1$, i.e., slow to moderate growth rates. For more rapidly growing populations, say, with 100 to 200% annual growth rate, i.e., r between 1 and 2, all solutions still tend to the carrying capacity K but now they oscillate on both sides of K as they converge to K.

As the percent growth rate crosses $r = 2$, motion becomes more complex. For r a little bigger than 2, say, 2.1, K is no longer a stable equilibrium. For, if the population level is a little below K one year, it will shoot well beyond K the following year into the region of negative growth, sending it back below K the next year. Now there is a population level below K to which the population returns every *two* years, so that the new stable "equilibrium" is a periodic motion that repeats its pattern every two years:

$$x_0 = a, \quad x_1 = b, \quad x_2 = a, \quad x_3 = b, \quad x_4 = a, \quad \text{etc.}$$

From any initial condition, each solution tends to this periodic solution as t gets larger.

As the growth rate r increases a bit more, this 2-year cycle becomes unstable and all solutions tend to a 4-year cycle:

$$x_0 = a, \quad x_1 = b, \quad x_2 = c, \quad x_3 = d, \quad x_4 = a, \quad x_5 = b, \quad \text{etc.}$$

These periodic phenomena are illustrated in the STELLA graph in Figure 6.5.

Finally, as r increases a bit further—to around $r = 2.5699456$—all these cyclic patterns disappear and the dynamics appear to cycle randomly, as pictured in Figure 6.6, for $r = 3$. In fact, one can prove that the dynamics in Figure 6.6 have the following properties—say, for $r = 3$:

Let x_0 be any be any initial population size in $(0, K)$, and let I be *any* very small interval of initial populations around x_0. Then,

1. there is an initial state y_0 in I whose corresponding "orbit" is a (regularly repeating) periodic cycle
2. there is another initial state z_0 in I whose corresponding orbit "fills up" the interval $(0, K)$ in the sense that, given any other point z^* in $(0, K)$, some iterate of z_0 under Equation 6.9 will agree with z^* to, say, one hundred decimal places
3. there is a w_0 in I whose corresponding orbit eventually strays far from the orbit through x_0. In other words, a small change in initial conditions will eventually lead to divergent growth patterns—a phenomenon known as "sensitive dependence on initial conditions"

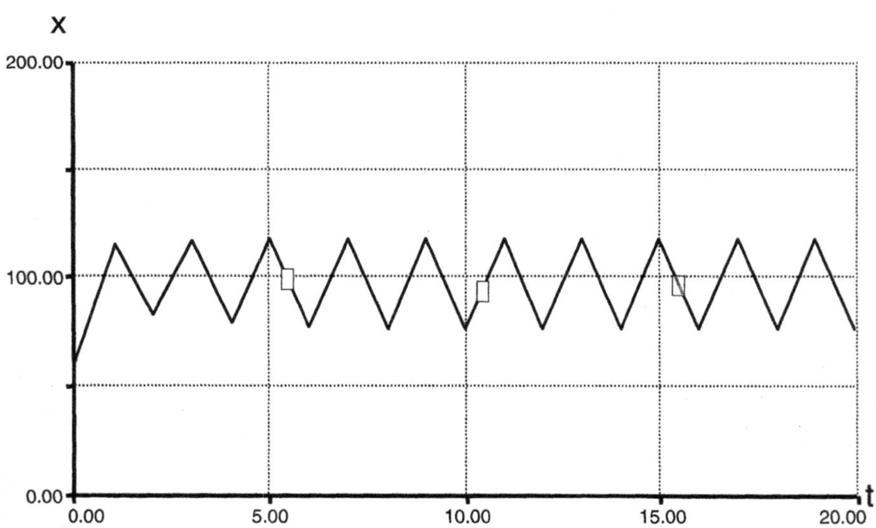

Figure 6.5 Periodic solution of discrete logistic Equation 6.9.

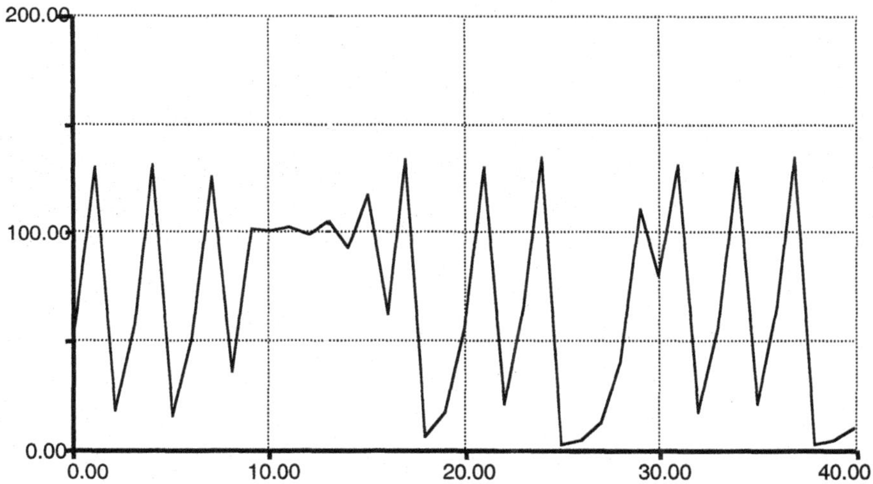

Figure 6.6 Chaotic solution of discrete logistic Equation 6.9.

A dynamic that has these properties is formally called **chaotic.** Chaotic dynamics appear unpredictable, and in a sense they are. However, at the same time, they have properties that make them amenable to analysis. A fish population with logistic growth Equation 6.9 that is chaotic is in trouble. For, by property 2, its time path will eventually take it dangerously close to $x = 0$ and extinction.

In summary, all continuous-time logistic growth and discrete-time logistic growth with $r < 2$ yield dynamics that always converge to the population carrying capacity K. However, for discrete logistic growth with $r > 2$, the dynamics can be periodic or chaotic or some combination of the two.

6.5 *More general growth rates*

This discussion of logistic growth was built on the assumption that the growth rate is a decreasing *linear* function of population size x. However, all these properties hold for *nonlinear* decreasing growth rates as well. Some, like the phase portrait in Figure 6.3, are still easy to demonstrate; others, like the complex behaviors of the discrete system, are much more difficult to verify.

Another useful growth assumption is that there is a population level K_1 below which the population will experience negative growth. One can think of K_1 as the minimum viable population level, below which the population cannot sustain itself. The corresponding percent growth rate $r(x)$ would have a graph as in Figure 6.7.

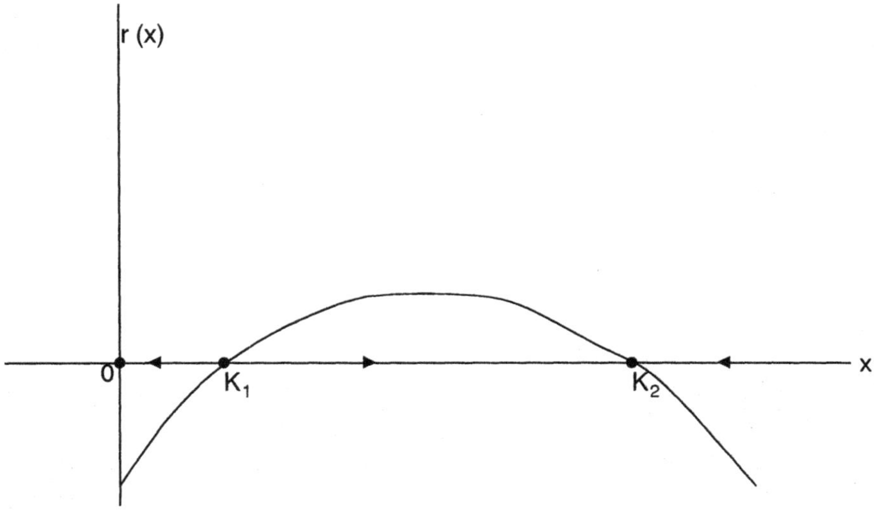

Figure 6.7 Growth curve for critical depensation.

Figure 6.8 Phase diagram for critical depensation.

The corresponding differential equation would be

$$\frac{dx}{dt} = x \cdot r(x). \tag{6.10}$$

with phase diagram as in Figure 6.8.

A simple example of Equation 6.10 is

$$\frac{dx}{dt} = rx\left(\frac{x}{K_1} - 1\right)\left(1 - \frac{x}{K}\right).$$

A population that has this kind of growth is said to exhibit **critical depensation.**

6.6 *Simple harvesting*

For simplicity of presentation, we focus next on a model of fisheries. Suppose there is a single species with population $x(t)$ at time t in a single location experiencing logistic growth. We next ask how different "harvesting" (i.e., fishing) patterns affect the growth of the population.

6.6.1 Constant harvest rate

The simplest pattern is a constant harvest rate h per unit of time. The underlying dynamic model is now:

$$\frac{dx}{dt} = rx\left(1 - \frac{x}{K}\right) - h. \tag{6.11}$$

Figure 6.9 shows the graph of each of the two terms in Equation 6.11.

When $A < x < B$, the extrinsic growth curve lies about the harvest rate h and the population grows. Otherwise, the extrinsic growth curve lies below h, the right-hand side of Equation 6.11 is negative, and the population declines. Figure 6.10 presents the corresponding phase diagram. All solutions with initial populations $x_0 > A$ tend to be level B as $t \to \infty$. However, if the initial population x_0 starts below A, then the population will die off: $x(t) \to 0$ as $t \to \infty$. The critical populations A and B are the zeros of the right-hand side of Equation 6.11. As the harvest rate h increases, so does the interval of

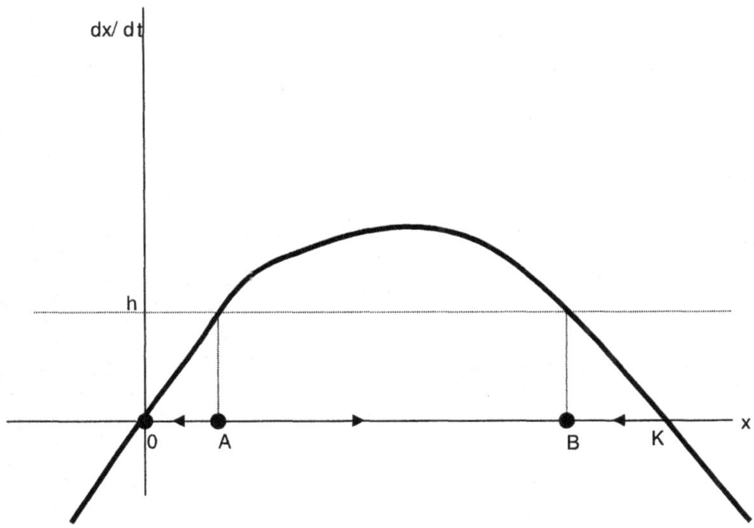

Figure 6.9 Graphs of functions on right-hand side of (11).

Figure 6.10 Phase diagram for Equation 6.11.

danger (0, A). The largest harvest rate h that crosses the intrinsic growth curve is $h = rK/4$; the maximum value of $rx(1 - (x/K))$. In fisheries literature $h = rK/4$ is called the **maximum sustainable yield (MSY).** If $h >$ MSY, the population will eventually collapse, for any initial x_0. (See Chapter 3.)

6.6.2 *Population-size-dependent harvest*

A constant harvest rate, independent of population size, is a somewhat unrealistic assumption. As the population dwindles, fish will generally be more difficult to harvest. The obvious way to change the model is to assume the harvest is an increasing function $h(x)$ of population size x: the bigger the population, the bigger the harvest. The simplest such function is a linear one: $h(x) = Ex$. In this case, the constant E is often called the **effort** (per unit of time). In some of our models, E is further broken down into (effort per boat) × (number of boats). In this case, "effort per boat" refers to the capabilities of the equipment on each boat. The corresponding differential equation is

$$\frac{dx}{dt} = rx\left(1 - \frac{x}{K}\right) - Ex. \tag{6.12}$$

Figure 6.11 summarizes the geometric solution of Equation 6.12.

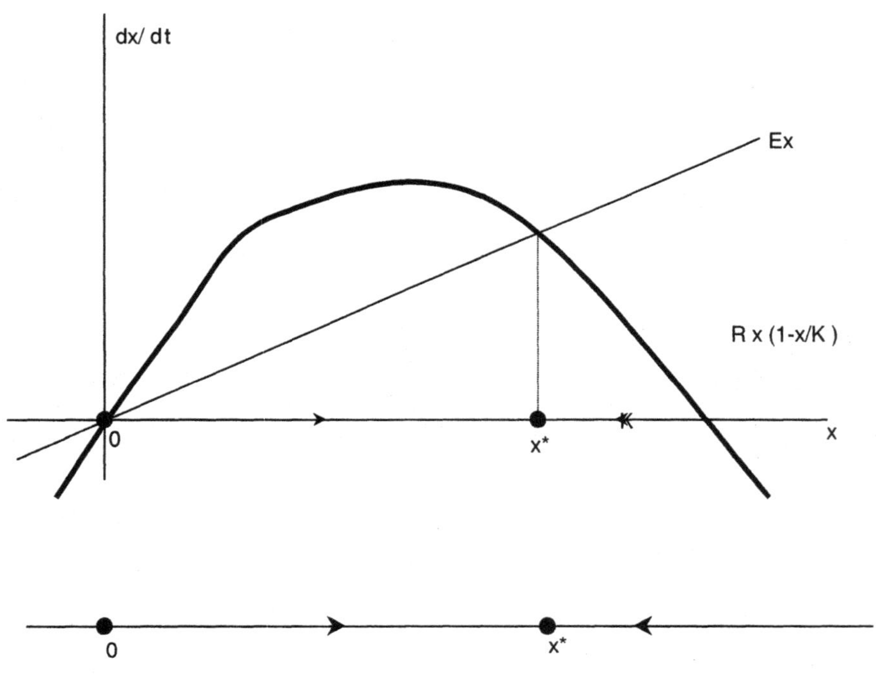

Figure 6.11 Phase diagram for Equation 6.12.

The two graphs in Figure 6.11 cross at:

$$x^* = K\left(1 - \frac{E}{r}\right) \tag{6.13}$$

which, of course, is also the zero of the right-hand side of Equation 6.12. All solutions of Equation 6.12 with $x_0 > 0$ tend to x^* of Equation 6.13. In fact, since the right-hand side of Equation 6.12 is a simple quadratic, equation Equation 6.12 can be analyzed just as the logistic equation Equation 6.8 was. The STELLA format of Equation 6.12 is:

$$x(t) = x(t - \Delta t) + r * x(t) * (1 - (x(t)/K)) - E * x(t),$$

with compartmental diagram as in Figure 6.12.

If the effort rate E is larger than the intrinsic growth rate r, then all solutions of Equation 6.12 tend to zero. Otherwise, the population will stabilize at non-zero level Equation 6.13.

6.7 Economic considerations

A simplistic approach to harvesting strategies had long suggested the **maximum sustainable yield (MSY)**—the maximum value of the growth curve (10)—as the optimal strategy. This approach has since been discredited because (1) it ignores the dependence of harvest on population size, and (2) it leads to a fragile situation for fishery survival. It also ignores economic motivation of the harvesters (Chapter 3). In this section we include the economic goals of the harvesters.

6.7.1 Constant price, sole owner

Suppose harvested fish sell for a market price p per unit. As suggested above, we factor effort E into the product of effort per boat q and number of boats b. (Some authors e.g., Clark, 1976 break down $E = qb$ as "catchability coefficient" q times "rate of fishery effort" b.) A harvester who uses b boats will expend total

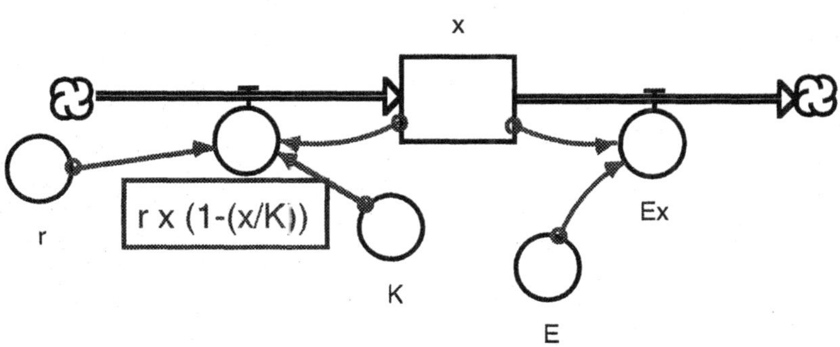

Figure 6.12 STELLA model for Equation 6.12.

effort $E = qb$. Assuming no other harvesters, the long-run equilibrium size of the fishery will be the x^* given by Equation 6.13 with long-run harvest rate:

$$h = Ex^* = EK\left(1 - \frac{E}{r}\right), \qquad E = qb. \tag{6.14}$$

The revenue from this harvest will be

$$R = ph = p \cdot EK\left(1 - \frac{E}{r}\right);$$

the cost will be $c \cdot b$, where c is the cost per boat. In terms of boats used b, the firm's profit function is

$$\Pi(b) = p \cdot q \cdot b \cdot K\left(1 - \frac{q \cdot b}{r}\right) - c \cdot b. \tag{6.15}$$

One can use either calculus or quadratic function analysis to compute the level of b that maximizes Equation 6.15:

$$b^* = \frac{r}{2q}\left(1 - \frac{c}{pqK}\right). \tag{6.16}$$

The resulting equilibrium fish population and harvest rates are

$$x^* = \frac{K}{2} + \frac{c}{2pq}, \tag{6.17}$$

and

$$h^* = qb^*x^*. \tag{6.18}$$

The situation in Equations 6.16 through 6.18 is the profit-maximizing activity, population size, and harvest rate when the fishery has a sole owner motivated by profit maximization.

6.7.2 Constant price, open access

The opposite situation to that of a single profit-maximizing owner of the fishery is the open-access fishery. In this case, no one has ownership rights to the fishery, and privately owned boats will keep entering as long as there are positive profits to be made. The long-run equilibrium in this case is sometimes called the **Gordon bionomic equilibrium** (Gordon, 1954; Clark, 1976). It is found by setting profits (Equation 6.15) equal to zero and solving for the corresponding x^*, h^*, and b^*, namely,

$$b^* = \frac{r}{q}\left(1 - \frac{c}{pqK}\right) \tag{6.19}$$

$$x^* = \frac{c}{pq} \tag{6.20}$$

$$h^* = Ex^* = \frac{cr}{pq}\left(1 - \frac{c}{pqK}\right). \tag{6.21}$$

Comparing Equations 6.16 through 6.18 with Equations 6.19 through 6.21, we see that the sole owner uses half as many boats as the open-access situation and that equilibrium population size is larger under the sole owner regime than under open access.

Figure 6.13 gives a graphical comparison of these two situations in (x, E) space. The negatively sloped line in Figure 6.13 is the set of equilibria (Equation 6.13); the vertical line is the zero industry-profit line (Equation 6.20); the hyperbolas are the level sets of the industry-profit function (Equation 6.15) with $E = qb$. Point A gives the sole-owner profit-maximizing

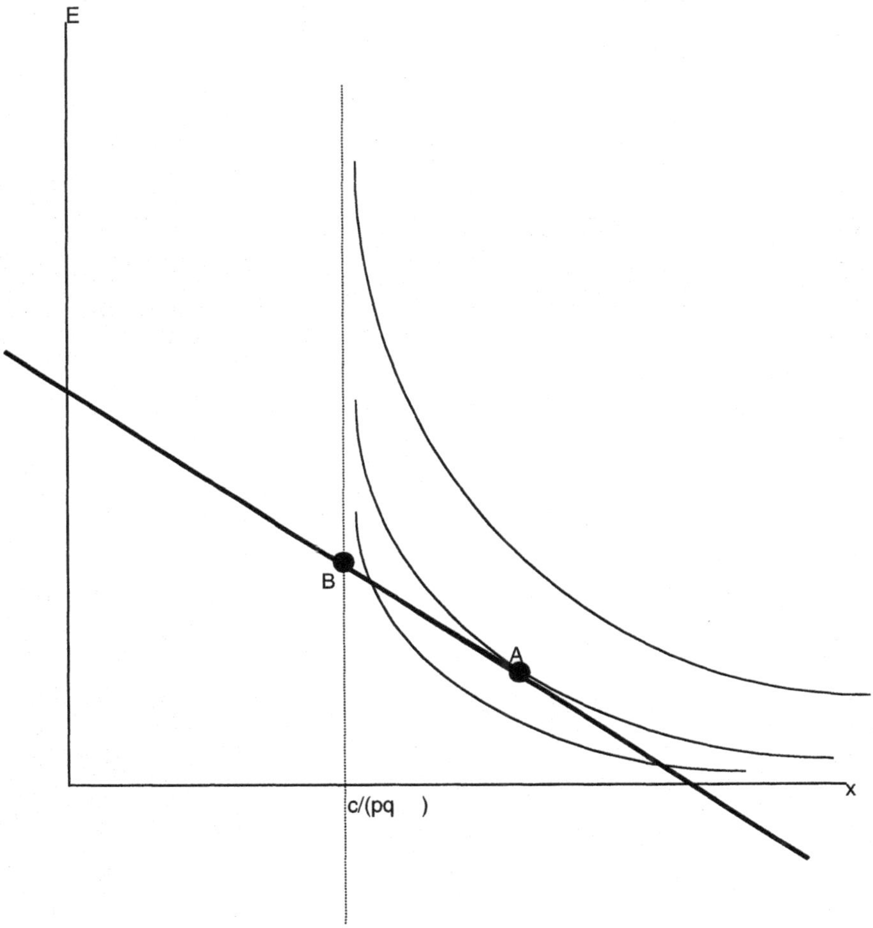

Figure 6.13 Level sets of harvest function (21).

(x^*, E^*) level of Equations 6.16 and 6.17. Point B gives the zero-profit open-access equilibrium of Equations 6.19 and 6.20. Figure 6.13 illustrates that the sole owner profit-maximizer equilibrium entails a larger equilibrium stock and lower effort level than does the open-access equilibrium. To compare actual harvest levels, one compares points A and B using the hyperbolic level sets of the harvest function $h = x \cdot E$.

6.7.3 Price-quantity dependence

The economics-oriented models in Sections 6.7.1 and 6.7.2 assumed that prices are independent of harvest amount—and vice versa. One can add realism to this model by adding a demand function $p = \alpha—\beta h$ to the above discussion. (If $\beta = 0$, we're back in the situation of the last two subsections). The profit function is now:

$$\Pi = (\alpha - \beta h) \cdot h - cb \tag{6.22}$$

where $h = EK\,(1 - (E/r))$ and $E = qb$. Putting all this together yields profit function:

$$\Pi(b) = (\alpha - bqbK\,(1 - (qb/r)))\,qbK\,(1 - (qb/r)) - cb. \tag{6.23}$$

Now both the sole owner case and the open-access market require the solution of a (more or less unsolvable) cubic equation in b to find optimal b^*, x^*, and h^*. We leave this situation as an exercise. The sole owner situation still entails fewer boats and a larger equilibrium population than does the open-access market.

6.8 Economic considerations: dynamic analyses

The economic considerations of the previous section require that the harvesters take a long-run equilibrium view in making their economic decisions. Although this approach leads to simple mathematical formulations, it does require a stretch of the imagination for the sole-owner and the open-access markets and is completely inconsistent with the very myopic nature of the open-access situation.

6.8.1 Perfect foresight: optimal control

A more realistic way of treating the sole owner scenario is to assume that the fish population follows a dynamic rule—not necessarily always at equilibrium—and that the owner tries to find a harvesting strategy over time that maximizes the present value of a (discounted) profit stream: choose effort level $E(t)$ to maximize:

$$\mathrm{PV}[E] = \int_0^{\infty} e^{-\delta t} \left[pE(t) \cdot x(t) - \frac{c \cdot E(t)}{q} \right] dt \tag{6.24}$$

where:

$$\frac{dx}{dt} = rx\left(1 - \frac{x}{K}\right) - E \cdot x \tag{6.25}$$

$$x \geq 0, \qquad 0 \leq E(t) \leq E_{max}, \qquad x(0) = x_0.$$

This is the approach that Clark (1976) emphasizes in his now classic text.

6.8.2 *Myopic adaptation: open access*

We will work with a more myopic approach that requires less omniscience on the part of the fishermen and owners—an approach that is more compatible with the STELLA dynamic compartmental approach. We will assume that the fishermen and fish start at some initial level of effort and population size, that the fish population obeys a logistic-like growth equation, and that fishermen keep making small adjustments to their fishing efforts that are consistent with the underlying market structure. We start with an open access model—in which harvesters respond to the sign of the total fishery profit. Fishermen will continue to add incremental effort (more boats) to the fishery as long as profits are positive. If incremental effort leads to a negative total profit, they will decrease their effort by this increment. A simple way to capture the rule that effort will increase over time if and only if total profits are positive is to assume that the incremental change in effort over time dE/dt is proportional to aggregate profit $\Pi(x, E)$. This groping movement ("tatonnement," in economics jargon) is captured by the system:

$$\frac{dx}{dt} = F(x) - Ex, \qquad \frac{dE}{dt} = \Pi(x, E), \tag{6.26}$$

where F is the intrinsic growth function for the fish. For simplicity, we will use the logistic function (8) for F and we will combine qb as E and just work with E. Further simplifying, we work here with perfectly elastic demand, i.e., price independent of quality, and leave the more realistic version as an exercise. In this framework, profit is:

$$\Pi = p \cdot h - cb = p \cdot Ex - \frac{cE}{q}.$$

So, system (26) becomes system (27) for the open-access situation:

$$\begin{aligned}\frac{dx}{dt} &= rx\left(1 - \frac{x}{K}\right) - Ex \\ \frac{dE}{dt} &= pEx - c\frac{E}{q}.\end{aligned} \tag{6.27}$$

The phase diagram for system 6.28 is given in Figure 6.14; the corresponding STELLA diagram is presented in Figure 6.15. With a bit of work, including the

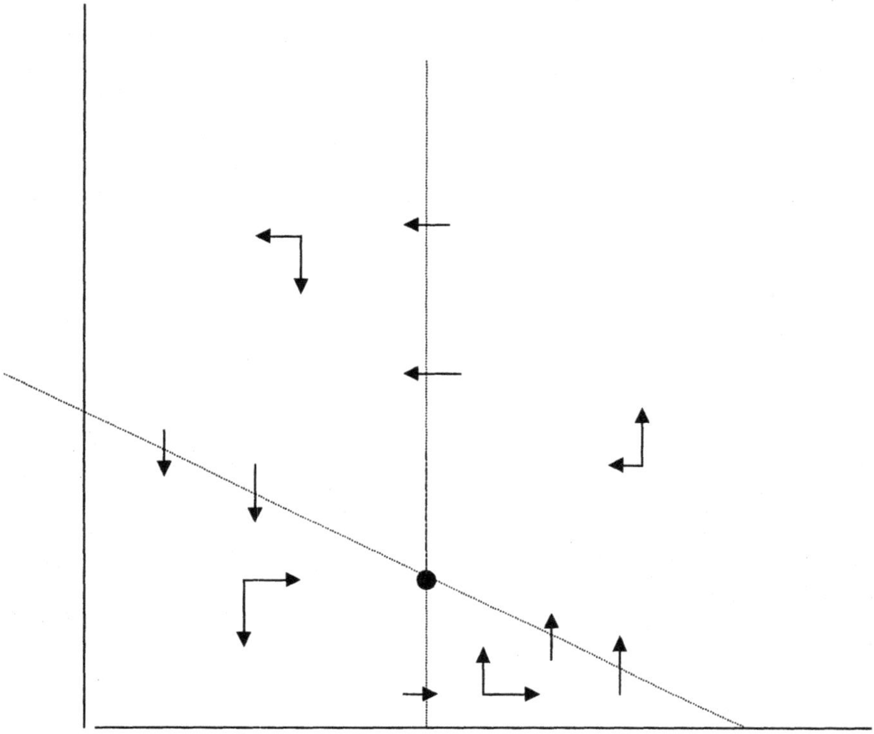

Figure 6.14 Phase diagram for system (27).

use of the Bendixson-DuLac Theorem (Clark, 1976) to rule out periodic orbits, one shows that all orbits of (27) tend to:

$$(x^*, E^*) = \left(\frac{c}{pq}, r\left(1 - \frac{c}{pqK}\right)\right). \tag{6.28}$$

This is precisely the equilibrium (19, 20) that we computed for the long-run-oriented open-access model in Section 6.7.2. Figure 6.16 shows the solution of the STELLA model portrayed in Figure 6.15 for parameter values: $c = 0.5$, $K = 100, p = 1, q = 0.01, r = 0.1$. As predicted in Equation 6.28, $x(t) \to 50$ and $E(t) \to 0.05$ as $t \to \infty$.

6.8.3 Including delays

In the real world, fishermen receive signals of the total industry profit with some delays and some uncertainty. It is difficult to include delays and uncertainty in the analytical models. (See Jacquez and Simon 2000 for some suggestions about adding delays into compartmental systems.) However, the STELLA program is a natural venue for dealing with both delays and uncer-

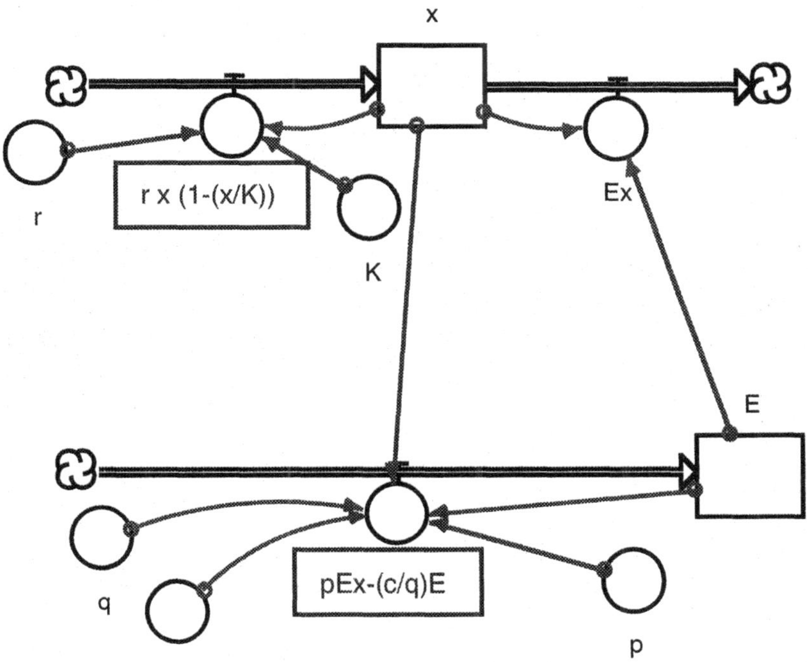

Figure 6.15 STELLA model for system (27).

tainty. Usually, delays add oscillations and some instability to the dynamics under study. While Figure 6.16 shows the STELLA-generated solutions of system 6.27 (without delays) tending to the equilibrium $x = 50$, $E = 0.05$, Figure 6.17 shows the solutions when a simple delay term is added to the second equation in (6.27). In this case, solutions starting very close to the above equilibrium oscillate away from it. See Chapter 3 for a more complete discussion of the inclusion of delays and uncertainty in fishery models.

6.8.4 Nanofish problem

The analytic solutions of system 6.27 oscillate toward the equilibrium $x = 50$, $E = 0.05$. However, some of these solutions get very close to the axes in the process. Analytically, this implies that the fish population or the number of boats used may fall below 10^{-10} at some time before it recovers to more reasonable values. STELLA, however, treats 10^{-10} as zero. If the fish population or the number of boats reaches this level in a STELLA run, STELLA computes it as zero and it stays at zero for all future time. For example, the effort level reached "STELLA zero" at the end of the graph in Figure 6.17—and remained at that level for all future times. This allowed the fish population to move back to carrying capacity. Since this situation does not exist for the

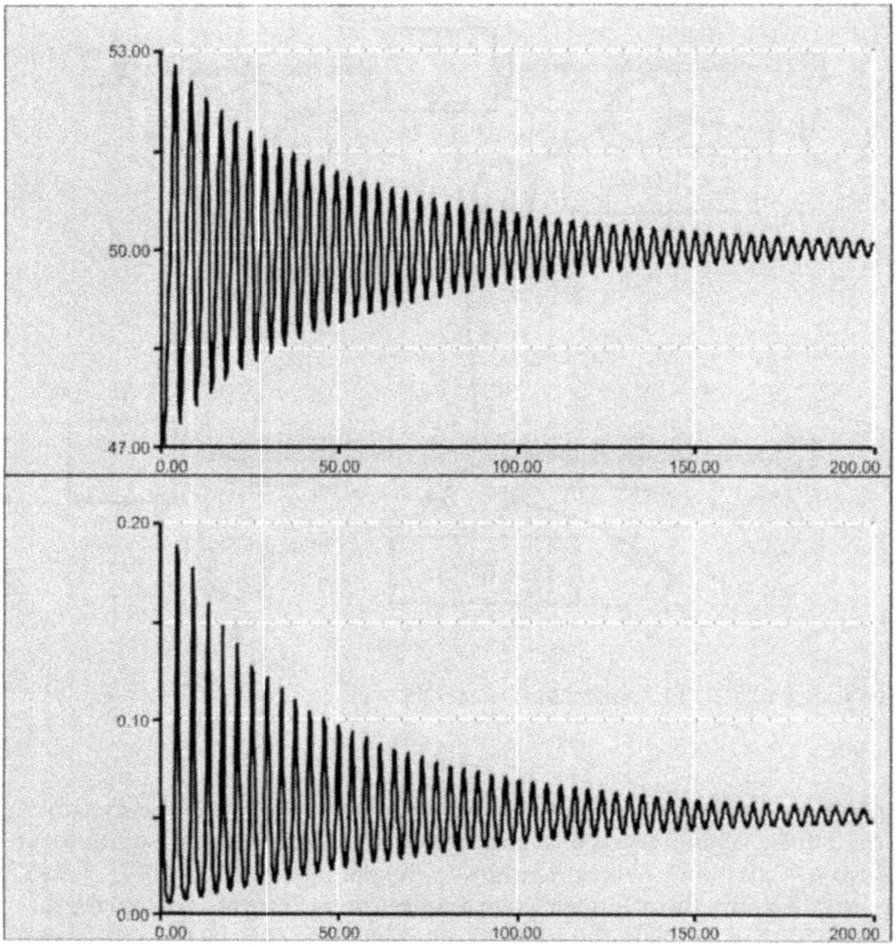

Figure 6.16 A STELLA solution for Equation 6.12.

analytic solution, this "nanofish" phenomenon is an important case where theory and computation diverge.

6.8.5 Myopic adaptation, sole owner

The sole-owner version of the myopic adaptive model has harvesters responding, not to the sign of the profit function, but to its dependence on the level of effort. If incremental effort leads to lower profit, fishermen will decrease their effort; if incremental effort leads to higher profit, they will increase effort by this increment. A simple way to capture the rule that one should increase effort if and only if such increased effort leads to increasing

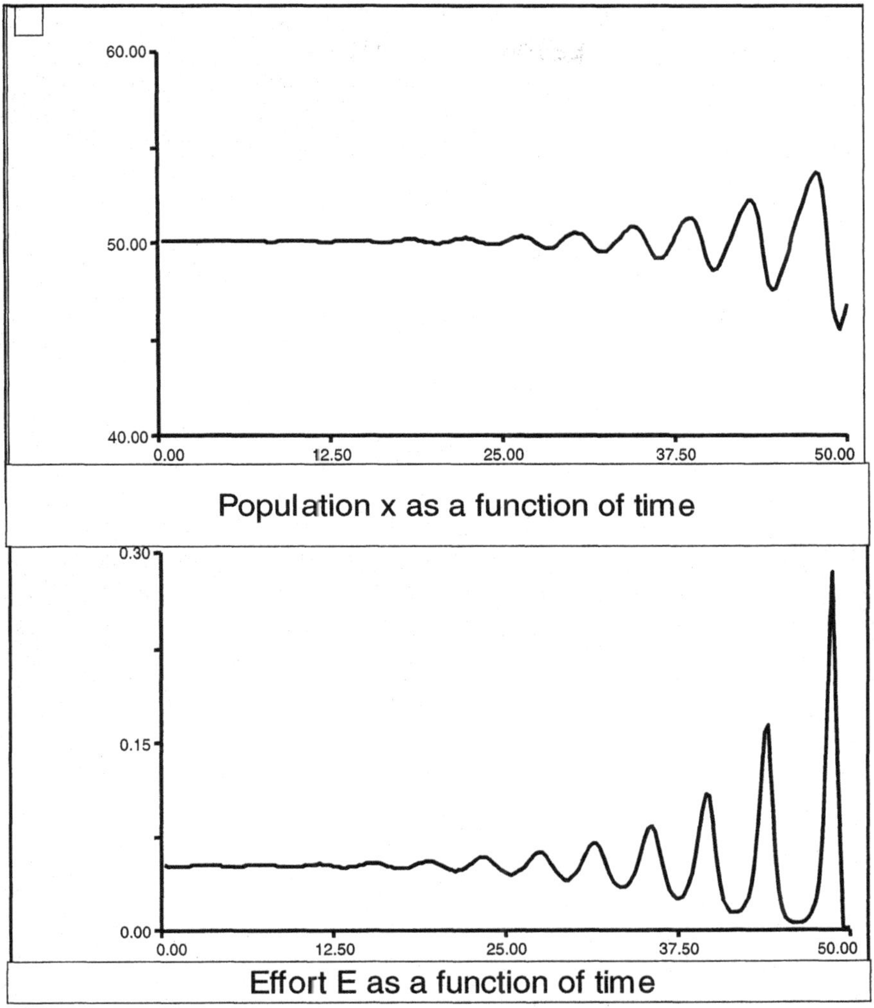

Figure 6.17 A STELLA solution for lagged version of system (27).

profit, is to assume that dE/dt is proportional to $\partial\Pi/\partial E$ In this case, $\partial\Pi/\partial E > 0$ implies increasing E leads to increasing profit Π, so we look for E to increase, i.e., $dE/dt > 0$, and vice versa for $\partial\Pi/\partial E < 0$. This tatonnement movement is captured by the system:

$$\frac{dx}{dt} = F(x) - Ex, \qquad \frac{dE}{dt} = \frac{\partial\Pi}{\partial E}(x, E), \tag{6.29}$$

where F is the intrinsic growth function for the fish. However, this approach does not work well in the (somewhat unrealistic) case of perfectly elastic demand, i.e., price independent of quantity harvested, since in this case the profit function is linear in E and $\partial\Pi/\partial E$ is independent of E. In the case of the more realistic demand function $h \to p - \beta h$, the system tends to the profit maximizing (x^*, E^*).

6.9 Multiple species

Our models have so far involved a single fish in a single location. The next complexity to include is the consideration of multiple species. We will assume that the various species compete for resources, but not that any one of them preys on any of the others. We will assume logistic growth for each species in the absence of the other species.

6.9.1 Gause competing species model

There are many ways to model how the species interact. One way is to assume that the growth of any one species directly impedes the growth of any other species. In the simplest case, this leads to the classical competing species model, studied analytically and experimentally by Gause (1935), which for two species look like

$$\begin{aligned}\frac{dx_1}{dt} &= x_1(r_1 - a_{11}x_1 - a_{12}x_2)\\ \frac{dx_2}{dt} &= x_2(r_2 - a_{22}x_2 - a_{21}x_1).\end{aligned} \tag{6.30}$$

In system 6.30, r_1 and r_2 are the natural growths rates of species 1 and species 2, respectively. Parameters a_{11} and a_{22} measure the intrinsic population bounds for each species; the carrying capacities of the two species are r_1/a_{11} and r_2/a_{22}, respectively. Parameters a_{12} and a_{21} measure how the presence of either species impinges on the growth of the other.

It can easily be shown (Simon and Blume, 1994; Clark, 1976) in this situation that every orbit tends to one of the following three equilibria:

$$A = \left(\frac{r_1}{a_{11}}, 0\right), \qquad \text{no species 2}$$

$$B = \left(0, \frac{r_2}{a_{22}}\right), \qquad \text{no species 1}$$

$$C = \left(\frac{r_1a_{22} - r_2a_{12}}{a_{11}a_{22} - a_{12}a_{21}}, \frac{a_{11}r_2 - a_{21}r_1}{a_{11}a_{22} - a_{12}a_{21}}\right), \qquad \text{coexistence.}$$

The table in 6.31 summarizes which parameter values in 6.30 lead to which of these outcomes. Figure 6.18 shows the phase diagram for the first case in this table.

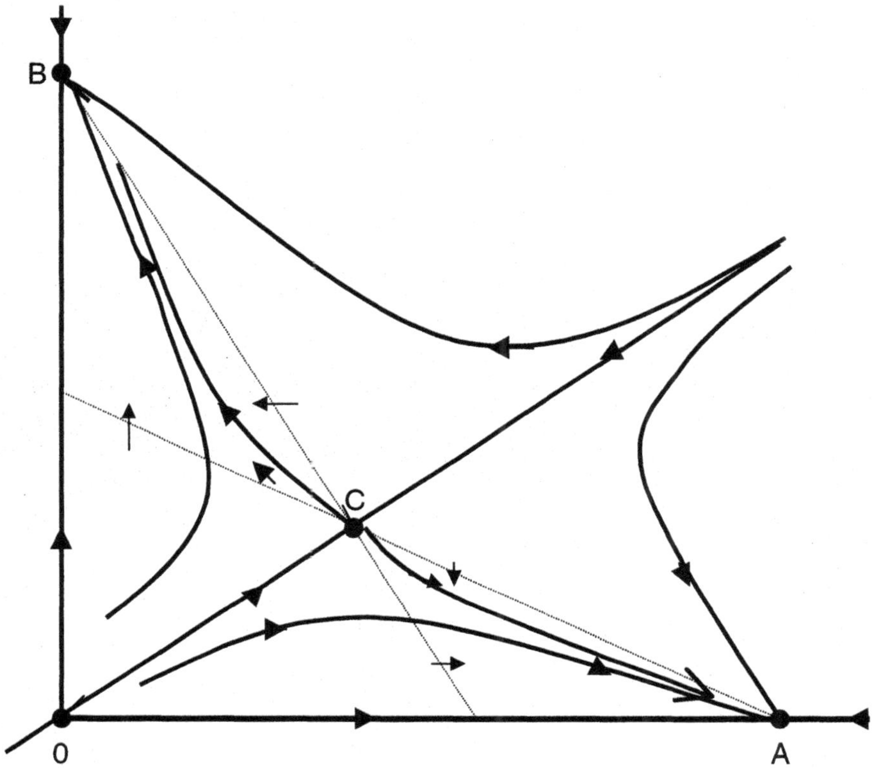

Figure 6.18 Phase diagram for realization 1 of system (30).

$$\begin{aligned}
&1.\ \frac{r_2}{a_{22}} > \frac{r_1}{a_{12}} \text{ and } \frac{r_1}{a_{11}} > \frac{r_2}{a_{21}} \Rightarrow A \text{ or } B\text{, depending on initial condition}\\
&2.\ \frac{r_2}{a_{22}} < \frac{r_1}{a_{12}} \text{ and } \frac{r_1}{a_{11}} < \frac{r_2}{a_{21}} \Rightarrow C\text{, coexistence}\\
&3.\ \frac{r_2}{a_{22}} > \frac{r_1}{a_{12}} \text{ and } \frac{r_1}{a_{11}} > \frac{r_2}{a_{21}} \Rightarrow B\text{, species 2 wins for all initial conditions}\\
&4.\ \frac{r_2}{a_{22}} < \frac{r_1}{a_{12}} \text{ and } \frac{r_1}{a_{11}} > \frac{r_2}{a_{21}} \Rightarrow A\text{, species 1 wins for all initial conditions.}
\end{aligned} \tag{6.31}$$

Note that three of the above four cases have only one species surviving. This illustrates the "principle" of competitive exclusion: usually only one species can occupy any ecological niche.

The corresponding dynamics for three competing species can be periodic (Zeeman, 1993), and for four competing species even chaotic (Arneado et al., 1982).

Suppose that the first species x^1 in system 6.30 is subject to harvesting, but not the second. For simplicity, we will assume constant effort E. Then, the first equation in 6.30 becomes

$$\begin{aligned}\frac{dx^1}{dt} &= x_1(r_1 - a_{11}x_1 - a_{12}x_2) - Ex_1\\ &= x_1((r_1 - E) - a_{11}x_1 - a_{12}x_2).\end{aligned}$$

So raising E lowers the effective r_1 in 6.31. One can imagine a scenario in which r_1 is large so that species 1 dominates, at first (case 4 in 6.31). Then, as E increases, the effective r_1 decreases so that case 4 becomes case 2 (coexistence), which later turns to case 3 (species 2 wins, species 1 dies out). According to Clark (Clark, 1976; Murphy, 1966), fisheries biologists propose such an explanation for the collapse of the Pacific sardine (*Sardinops caerula*) fishery in the late 1940s and early 1950s. The sardines had been in competition with a population of anchovies (*Engaulis mordax*), which is now dominant and is inhibiting the recovery of the sardines.

6.9.2 *Constant total fishery carrying capacity*

Wilson et al. (1991) argues that when there are multiple subpopulations in a given fishery, energy considerations suggest that carrying capacity constraints apply more naturally to the whole fishery and not to individual species. In other words, the biomass of the community as a whole is relatively stable. To model this approach, we will assume a logistic growth curve for each subpopulation i with corresponding carrying capacity K_i, but modified by the factor $1 - (x_1 + \cdots + x_n)/K_0$ to capture the effect of the community carrying capacity K_0. The fishery dynamics becomes

$$\begin{aligned}\frac{dx_1}{dt} &= r_1x_1\left(1 - \frac{x_1}{K_1}\right)\left(1 - \frac{x_1 + \cdots + x_n}{K_0}\right)\\ &\vdots \qquad \vdots \qquad \vdots \qquad\qquad \vdots\\ \frac{dx_n}{dt} &= r_nx_n\left(1 - \frac{x_n}{K_n}\right)\left(1 - \frac{x_1 + \cdots + x_n}{K_0}\right).\end{aligned} \tag{6.32}$$

In this system the different subpopulations interact only through their impacts on the community carrying capacity.

System 6.32 exhibits reasonably straightforward dynamics. We will assume throughout that each subpopulation i starts below its carrying capacity K_i. If $K_0 > K_1 + \cdots + K_n$, then each subpopulation x_i tends monotonically to its carrying capacity K_i as $t \to \infty$. If $K_0 < K_1 + \cdots + K_n$, then $(x_1(t), \ldots, x_n(t))$ tends to an equilibrium $\mathbf{x}^* = (x_1^*, \ldots, x_n^*)$ that satisfies $\Sigma_i x_i^* = K_0$, but varies continuously with $(x_1(0), \ldots, x_n(0))$; small changes in initial conditions lead to small changes in the long run equilibrium $\mathbf{x}^*$. During the dynamic process, $x_1(t) + \cdots + x_n(t)$ increases (or decreases) monotonically to K_0. In particular,

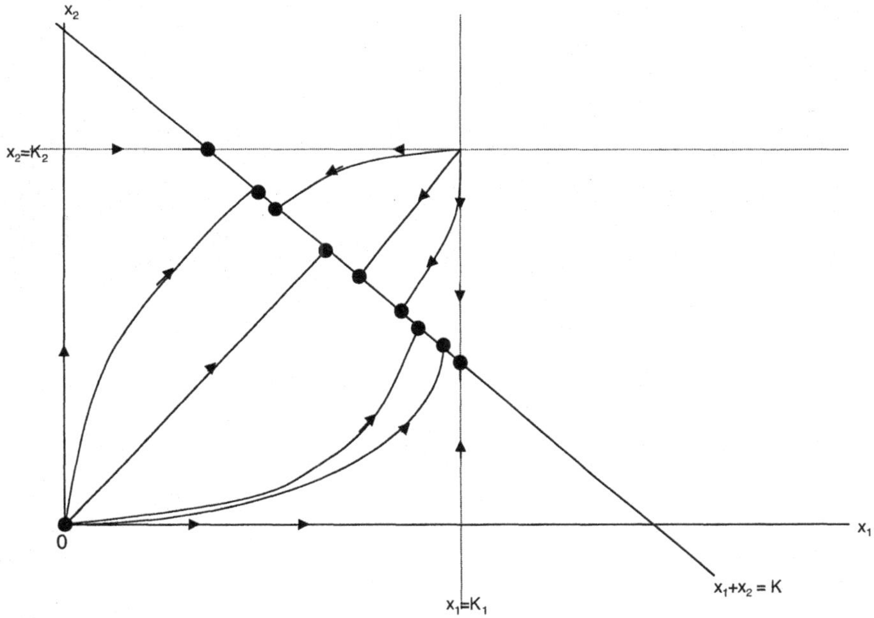

Figure 6.19 Phase diagram of system (32) for n=2.

there are no periodic solutions. Figure 6.19 presents the phase diagram for a two-subpopulation model of 6.32 with $K_0 < K_1 + K_2$ and suggests the proof of the assertions in this paragraph.

6.10 Age structure

Thus far we have ignored age and size differences within each fish population. However, for most species not every age or size is desirable or legal for harvesting. We bring age structure into our models in this section, focusing first on a single population.

6.10.1 Single population

Suppose that the maximum age at death of a member in the population under study is N years. Divide the population into $N + 1$ cohorts by age; a fish is in cohort i if it was born between i and $i + 1$ years ago. Let y_i denote the size of age cohort i, $i = 0, 1, \ldots, N$. To complete the age dynamics, we need to know (1) what proportion s_i of those in cohort i survive to make it to cohort $i + 1$, and (2) the yearly average number of births b_i by a member of cohort i. To keep the accounting simple, we usually only keep track of the female members in each age cohort. Of course, $s_i = 1 - m_i$, where m_i is the average annual mortality rate for age cohort i.

If we assume that the s_i and b_i terms are constants independent of cohort size, we have a linear dynamic leading to exponential growth:

$$\begin{aligned} y_0(t+1) &= b_0 y_0(t) + \ldots b_{N} y_N(t) \\ y_1(t+1) &= s_0 y_0(t) \\ \vdots \quad & \quad \vdots \\ y_N(t+1) &= s_{N-1} y_{N-1}(t). \end{aligned} \tag{6.33}$$

The coefficient matrix:

$$L = \begin{pmatrix} b_0 & b_1 & \cdots & b_{N-1} & b_n \\ s_0 & 0 & \cdots & 0 & 0 \\ \vdots & \vdots & \ddots & \vdots & \vdots \\ 0 & 0 & \cdots & s_{N-1} & 0 \end{pmatrix}$$

of system 6.33 is called a **Leslie matrix.** Its largest eigenvalue gives the long-run population growth rate. The corresponding eigenvector gives the long-run distribution of ages in the population. Caswell (1989) provides a complete discussion of linear population models and their analysis.

To make this model more appropriate for fish populations, we assume that only certain age cohorts can lay eggs (spawn) and that the average spawner in cohort i lays e_i eggs that successfully hatch each year. One can bring into this model limits to growth, i.e., the carrying capacity effect, in many ways. For example, one can assume that the mortality rate for newborns $m_0 = m_0(Y)$ is an increasing function of the size of the *entire* population $Y = y_0 + \cdots + y_N$ and that $m_0(Y) = 1$ for all $Y \geq K$ for some population carrying capacity K. An example of such a function is

$$m_0(Y) = \min\left\{\left(\frac{Y}{K}\right)^a, 1\right\},$$

for some constant a, $a > 1$ so that m_0 is a convex function on $[0, K]$.

If we take $N = 4$ and cohorts 2, 3, and 4 to be the spawners with average individual annual egg clusters of size e_2, e_3, and e_4, respectively, then the population dynamic becomes

$$\begin{aligned} y_0(t+1) &= e_2 y_2(t) + e_3 y_3(t) + e_4 y_4(t) \\ y_1(t+1) &= (1 - m_0(Y(t)))y_0(t) \\ y_2(t+1) &= (1 - m_1)y_1(t) \\ y_3(t+1) &= (1 - m_2)y_2(t) \\ y_4(t+1) &= (1 - m_3)y_3(t) \end{aligned} \tag{6.34}$$

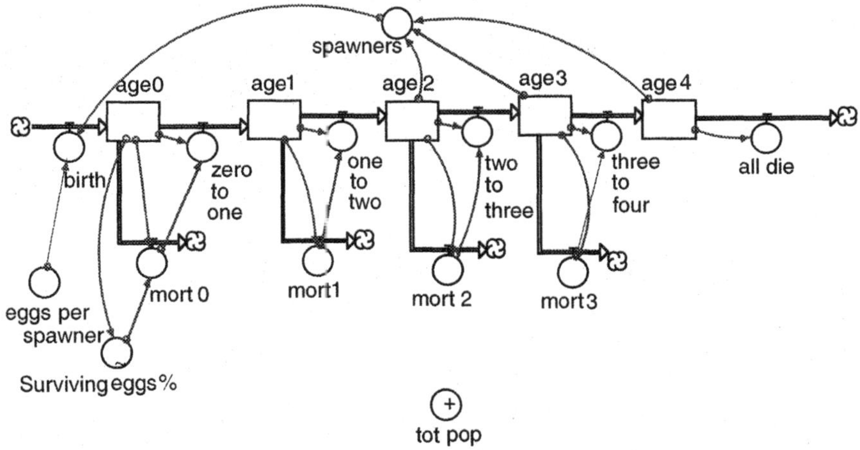

Figure 6.20 STELLA model for system (34).

Implicitly, $m_4 = 1$. The STELLA diagram for system 6.34 is presented in Figure 6.20.

One may want to use a smaller unit of time Δt than a year, especially if most of the cohort 0 mortality occurs in the first Δt of a year. One would then multiply the number of cohorts by $1/\Delta t$, with corresponding increases in the number of b_i terms and s_i terms (or m_is) that one must consider.

The differential equation version of system 6.34 is

$$
\begin{aligned}
\frac{dy_0}{dt} &= e_2 y_2 + e_3 y_3 + e_4 y_4 - y_0 \\
\frac{dy_1}{dt} &= (1 - m_0(Y))y_0 - y_1 \\
\frac{dy_2}{dt} &= (1 - m_1)y_1 - y_2 \\
\frac{dy_3}{dt} &= (1 - m_2)y_2 - y_3 \\
\frac{dy_4}{dt} &= (1 - m_3)y_3 - y_4.
\end{aligned}
\tag{6.35}
$$

The intermediate step between the last equation in the discrete-time system (6.34) and the continuous-time system (6.35) is

$$y_4(t + \Delta t) = (1 - m_3)y_3(t) \cdot \Delta t + (1 - \Delta t)y_4(t). \tag{6.36}$$

The first term on the right side of 6.36 indicates that if we divide each year into $1/\Delta t$ intervals, only those in the last interval move to the next age cohort; the second term on the right indicates that those in the first $(1/\Delta t) - 1$

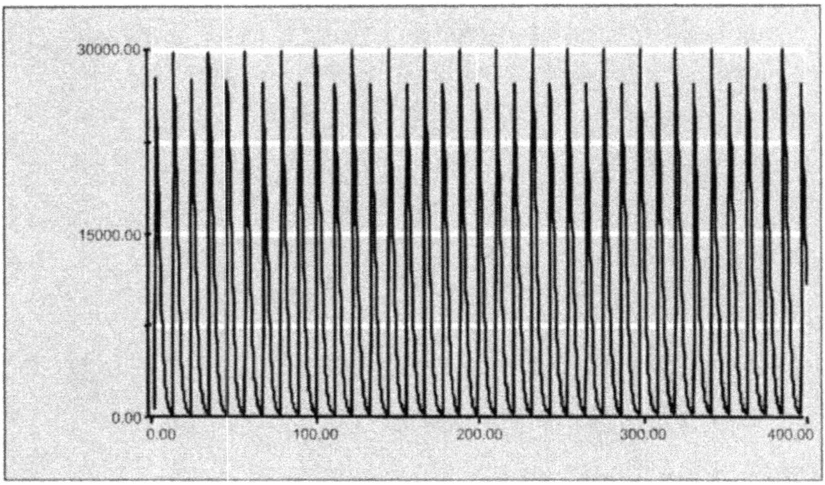

Figure 6.21 A solution of STELLA model for system (34).

intervals in cohort 4 stay in cohort 4. Subtract $y_4(t)$ from both sides of 6.36 and divide everything by Δt:

$$\frac{y_4(t+\Delta t) - y_4(t)}{\Delta t} = (1 - m_3)y_3(t) - y_4(t).$$

Now, let Δt tend to 0. It is often easier to compute the (nonzero) equilibrium age distribution and its stability for system 6.35 than it is for system 6.34. One often finds that this equilibrium is unstable and that STELLA solutions of 6.35—and of 6.34—tend to oscillate around this equilibrium, as Figure 6.21 indicates.

6.10.2 Multiple populations

One can also consider *multiple* age-structured populations and have them interact with each other in various ways. A simple interaction device is to have the cohort 0 mortality m_0^i of each subpopulation i be an increasing function of the sum of *all* the subpopulation sizes $m_0^i(\Sigma_j Y^j)$, with $m_0^i = 1$ for all $\Sigma_j Y^j \geq K_0$, a metapopulation carrying capacity.

6.11 Multiple locations and multiple harvesters

In most ocean and many inland fisheries, there are a number of different harvesters and each has a choice over a number of fisheries. In this section we will suppose, for the sake of simplicity, that there are $N = 3$ different fisheries to choose from; the techniques discussed work equally well for general N. For each fishery i, we define the following variables:

$$
\begin{aligned}
x_i &= \text{size of fish population in fishery } i \\
r_i, K_i &= \text{intrinsic growth rate and carrying capacity in fishery } i \\
p_i &= \text{price of fish in fishery } i \\
b_i &= \text{number of boats fishing in fishery } i \\
c_i &= \text{cost per boat in fishery } i \\
h_i &= \text{harvest rate in fishery } i \\
\Pi_i &= \text{total profits in fishery } i \\
\pi_i = \Pi_i/b_i &= \text{profit per boat in fishery } i
\end{aligned}
$$

We assume that p_i and h_i are related by an (inverse) demand function, for example, $p_i = \alpha_i - \beta_i h_i$ (linear), $p_i = \alpha_i h_i^{-\beta_i}$ (constant elasticity), or p_i constant (perfectly elastic). Since we will not be carrying out many calculations in this section, we will not further specify which demand function holds in which fishery. When we do such calculations, we will assume perfectly elastic demand.

We assume that the harvest rate h_i is an increasing function of the number of boats b_i and of the fish population x_i, for example, a Cobb–Douglas production function:

$$h_i(b_i, x_i) = a_i \cdot b_i^{\gamma_i} \cdot x_i^{\nu_i}. \tag{6.37}$$

In this case, $0 < \gamma_i \leq 1$ and $0 < \nu_i \leq 1$. The exponent ν_i is a measure of the patchiness of fish distribution in fishery i. The distribution of a non-schooling fish like flounder may be fairly uniform; in this case $\nu_i = 1$ would be most appropriate. For a patchily distributed fish, like herring, $\nu_i \ll 1$.

The profit Π_i in fishery i equals revenues minus costs:

$$\Pi_i = p_i(h_i)h_i - c_i b_i, \qquad \text{where } h_i = h_i(b_i, x_i).$$

The fish dynamic in fishery i is the usual logistic dynamic:

$$\frac{dx_i}{dt} = r_i x_i\left(1 - \frac{x_i}{K_i}\right) - h_i(b_i, x_i). \tag{6.38}$$

(For the sake of simplicity, we are only writing down the continuous-time models; our discussion will include both continuous-time and discrete-time formulations.) If the fish can move easily from one fishery to the others, we might use the dynamic:

$$\frac{dx_i}{dt} = r_i x_i\left(1 - \frac{x_i}{K_i}\right)\left(1 - \frac{x_1 + x_2 + x_3}{K_0}\right) - h_i(b_i, x_i),$$

where K_0 is the metapopulation carrying capacity.

To complete the model, we need to include the economic motives of the fishermen.

6.11.1 Profit-maximizing sole owner

If all the boats in all fisheries are owned by a single, profit-maximizing owner, then the owner will want to choose (b_1, b_2, b_3), the number of boats in each fishery, so as to maximize *total* profits $\Pi_1 + \Pi_2 + \Pi_3$. Of course, these choices will depend on the size of the fish subpopulations, which in turn depend on the b_i values. We will assume decisions are made incrementally. In the discrete-time case, the owner will look at how profits responded to increasing or decreasing the number of boats in the past few periods. If, for example, increasing the number of boats in fishery 1 led to decreased profit, the owner will decrease the number of boats in fishery 1 next period. In the continuous-time model, the owner will choose b_i so that db_i/dt is proportional to $\partial(\Pi_1 + \Pi_2 + \Pi_3)/\partial b_i = \partial\Pi_i/b_i$, so that if increasing b_i raises profit, the owner will raise b_i, as we discussed above. The full six-equation dynamic model is:

$$
\begin{aligned}
\frac{dx_1}{dt} &= r_1x_1\left(1 - \frac{x_1}{K_1}\right) - a_1b_1^{\gamma_1}x_1^{\nu_1}\\
\frac{dx_2}{dt} &= r_2x_2\left(1 - \frac{x_2}{K_2}\right) - a_2b_2^{\gamma_2}x_2^{\nu_2}\\
\frac{dx_3}{dt} &= r_3x_3\left(1 - \frac{x_3}{K_3}\right) - a_3b_3^{\gamma_3}x_3^{\nu_3}\\
\frac{db_1}{dt} &= \kappa_1\frac{\partial\Pi}{\partial b_1} = \kappa_1(p_1a_1\gamma_1b_1^{\gamma_1-1}x_1^{\nu_1} - c_1)\\
\frac{db_2}{dt} &= \kappa_2\frac{\partial\Pi}{\partial b_2} = \kappa_2(p_2a_2\gamma_2b_2^{\gamma_2-1}x_2^{\nu_2} - c_2)\\
\frac{db_3}{dt} &= \kappa_3\frac{\partial\Pi}{\partial b_3} = \kappa_3(p_3a_3\gamma_3b_3^{\gamma_3-1}x_3^{\nu_3} - c_3),
\end{aligned}
\tag{6.39}
$$

where for simplicity we use perfectly inelastic demand and the production functions from (37). One would naturally want to include a time delay in this system, realizing it takes some time for profit information to affect boat use decisions.

6.11.2 Open access

As we discussed above, in an open-access economy, harvesters move into each fishery as long as there are positive (economic) profits to made there. In the long–run equilibrium, the profits in each fishery will be zero. In the short–run discrete dynamic, the number b_i of boats in fishery i will increase by one at time $t + 1$ if the boats in fishery i made positive profits last period. In the continuous-time model a natural dynamic is that db_i/dt is proportional to Π_i, or better yet, to $\pi_i = \Pi_i/b_i$ the profits per boat in fishery i. In this case, we would replace the last three equations in (6.39) by the following three equations:

$$
\frac{db_1}{dt} = \kappa_1\frac{\Pi_1(b_1)}{b_1} = \kappa_1(p_1a_1b_1^{\gamma_1-1}x_1^{\nu_1} - c_1)
$$

$$\frac{db_2}{dt} = \kappa_2 \frac{\Pi_2(b_2)}{b_2} = \kappa_2(p_2 a_2 b_2^{\gamma_2 - 1}{}_2 x_2^{\nu_2} - c_2)$$
$$\frac{db_3}{dt} = \kappa_3 \frac{\Pi_3(b_3)}{b_3} = \kappa_3(p_3 a_3 b_3^{\gamma_3 - 1}{}_3 x_3^{\nu_3} - c_3)$$

6.11.3 *Myopic profit adjustment*

In the above cases—sole owner and open access, the number of boats in each fishery is endogenous. Finally, we consider a case between these two. Suppose that each boat is privately owned and that these owners respond to profit signals in deciding which fishery to fish; that is, they tend to move to a fishery that yields higher profit than their current fishery, if there is one. In the discrete-time model, each boat studies the time series of the profits per boat in each fishery and, with some probability, moves to a more profitable fishery. In the continuous-time case, a natural candidate for the db_i/dt dynamic is

$$\frac{db_1}{dt} = \lambda_{12}(\pi_1(b_1) - \pi_2(b_2)) + \lambda_{13}(\pi_1(b_1) - \pi_3(b_3))$$
$$\frac{db_2}{dt} = \lambda_{12}(\pi_2(b_2) - \pi_1(b_1)) + \lambda_{23}(\pi_2(b_2) - \pi_3(b_3))$$
$$\frac{db_1}{dt} = \lambda_{13}(\pi_3(b_3) - \pi_1(b_1)) + \lambda_{23}(\pi_3(b_3) - \pi_2(b_2))$$

In these equations the λ_{ij} terms are measures of how easy it is to move between fishery *i* and fishery *j*. Note that $\Sigma_i db_i/dt = 0$, so that the total number of boats remains constant.

6.11.4 *Fish migration*

If the fisheries are connected, we can also include cross-migration terms for the fish, just as we did for the fishermen in the open-access section of multiple locations and multiple harvest. In this case, fish would tend to move from the more crowded to the less crowded fisheries. See Chapter 3 for further discussion of this issue.

$$\frac{dx_1}{dt} = r_1 x_1 \left(1 - \frac{x_1}{K_1}\right)\left(1 - \frac{x_1 + x_2 + x_3}{K_0}\right) - \mu_{12}(x_1 - x_2) - \mu_{13}(x_1 - x_3)$$
$$\frac{dx_2}{dt} = r_2 x_2 \left(1 - \frac{x_2}{K_2}\right)\left(1 - \frac{x_1 + x_2 + x_3}{K_0}\right) - \mu_{12}(x_2 - x_1) - \mu_{23}(x_2 - x_3)$$
$$\frac{dx_3}{dt} = r_3 x_3 \left(1 - \frac{x_3}{K_3}\right)\left(1 - \frac{x_1 + x_2 + x_3}{K_0}\right) - \mu_{13}(x_3 - x_1) - \mu_{23}(x_3 - x_2).$$

6.12 *Comparing solution strategies*

In this chapter we have presented an introduction to mathematical models of fisheries ecology, with an emphasis on the simple models that can be solved analytically. We have also indicated how one might use the program STELLA to study these and more complex models. In general, it is a good idea to start with the simplest models of any phenomenon and understand their structures and solutions before advancing to more complex models with more parameters and more variables.

There are real advantages to being able to compute an analytic solution of a model. Such analytical results are general theorems that hold for *all* values of the parameters and *all* initial population levels. On the other hand, strictly speaking, results of STELLA runs yield true propositions only for those parameter values and initial conditions actually implemented on the computer runs. For example, one could implement on STELLA many runs of the two-species competing species model 6.30, using many different values for the intrinsic growth rates and carrying capacities. All these runs will tend to some equilibrium. Yet, it takes a mathematical proof to assert that *every* solution of system 6.30 tends to an equilibrium—for all parameter values and for all initial conditions. Furthermore, the mathematical analysis gives an analytic expression for these equilibria and specifically indicates which equilibrium occurs for which choice of parameter values, as indicated in 6.31. Mathematical analysis can also demonstrate solution properties not evident in typical STELLA runs and can yield nonevident properties of the solutions that STELLA does find (for example, properties 1 and 2 for the logistic Equation 6.9 with $r = 3$).

STELLA certainly has its advantages. It is much easier to gain insights on a dynamical system by running STELLA on it than it is by carrying out mathematical analysis on it. More importantly, STELLA runs can give valuable insights into models that are too large or too complex to be amenable to mathematical analysis. This includes models with the realistic possibilities of lags or uncertainty in the dynamical system's response. Certainly, most realistic fishery models fit into this category. Rarely does the equation under study perfectly mirror the real-world situation. With STELLA one can easily try perturbations of the system under study to test for robustness and resilience.

section three

Models and empirical studies in the real world

chapter seven

Simulation of the market for Maine lobster

James Wilson

Contents

The relationship of market institutions and resource sustainability is an area of great interest but one that is rarely investigated beyond statistical estimates of price changes. In this chapter we describe a model that simulates the institutions that govern the decision behavior of agents (traders) in the market for Maine lobster. The purpose of the model was to understand how factors such as inventory capacity and the timing and volume of landings might interact with possible new harvesting rules to affect prices and ultimately the efficacy of the harvesting rules. The chapter describes how we were able to mix qualitative information obtained from interviews with dealers (traders) and quantitative information obtained from official data sources to quantify model structure and estimate changes in prices, inventories, and so on. We are not aware of any similar work in the literature; consequently, what we

1-5667-0389-1/01/$0.00+$.50

present here should be taken as a preliminary exploration of a new and possibly useful methodology. For the same reason, our description pays unusual attention to the actual process we went through in the development of the model.

In a recent article in *Scientific American,*[1] the economist Donald McCloskey outlines three phases in the development of a science—the philosophical, the statistical, and the Babylonian. The phases are differentiated principally by the costs of measurement and calculation, with the Babylonian phase representing a brute-force emphasis on cheap calculations. As will become apparent to the reader, the approach we describe here is most definitely of the Babylonian type.

7.1 Industry and market background

Maine lobsters are caught from Newfoundland south to Cape Hatteras, with the greater part of the landings occurring in Maine and the Canadian Maritimes. Differences in Canadian and U.S. regulations effectively create two seasonally complementary markets. U.S. harvests are not limited to regulated seasons and tend to concentrate in the summer and early fall. Canadian regulators have imposed harvesting seasons that tend to occur at times when U.S. supplies are relatively low. Despite the complementary nature of Canadian and U.S. landings, prices tend to vary by a factor of two or more from lows, usually in September, to highs, usually in February or March. The industry engages in months-long inventorying of live lobsters, which are used to supply the market at times when winter weather and ice usually prevent reliance on freshly landed lobsters and, of course, lead to high prices. Canadian regulators have required, in some areas, different minimum sizes of capture, which creates a canned and frozen market that serves segments of the market not open to live lobsters.

7.2 The management/regulatory problem

Unlike many of the fisheries of the northwest Atlantic, lobster populations and harvests have grown steadily since the late 1970s, possibly as a compensating response to the depleted state of other fisheries. In spite of this history, models used by the National Marine Fisheries Service (NMFS) indicated a severely overfished population throughout the 1970s, 1980s, and 1990s. As a result, the New England Fisheries Management Council contemplated a new regulatory regime that would require reductions in fishing effort of 20 to 50% depending on the region within the overall fishery. The council had established industry advisory groups, termed effort management teams (EMTs), and had given these groups the option of choosing their preferred approach

[1] Donald McCloskey, Computation Outstrips Analysis, *Scientific American*, July 1995, p. 26.

to effort reduction. Among the alternatives open to and considered by the EMTs were individual transferable quotas, seasonal closures, limits on the amount of fishing gear, and a variety of other possibilities.

The EMTs were particularly interested in how any alternative they might choose would alter, not only the economics of harvesting, but also the behavior of the market, including prices. Not incidentally they were also interested in how changes in the market would feed back to the harvesting sector and influence the implementation and economics of any chosen method of fishing effort reduction.

In order to answer questions of this sort, the NMFS contracted with investigators at four New England universities (Maine, New Hampshire, Massachusetts (Dartmouth) and Rhode Island) to construct a large simulator of the entire industry—harvesting and market. The market simulator reported on here is part of that larger simulation effort.

7.3 Building the market simulator

Given the regulatory alternatives under consideration by the EMTs, the strong dependence of the industry on diverse but seasonal markets, the problems with price volatility and the importance of pounding,[2] we decided early in the process that the usual estimate of demand elasticity would be insufficient for the industry's purposes. We were also worried that regulatory changes might be so significant that the ability to statistically extrapolate from the history of the market might be strongly impaired. Almost all of the possible regulatory changes were expected to affect the volume and timing of landings. Consequently, it was important that for each regulatory alternative analyzed, we be able to give some guidance with regard to the complex interplay between the timing and volume of landings and imports, pounding and pound capacity, and prices. For this reason we decided that we would attempt to construct a simulation model of the market—one that included the factors critical to the regulatory decision. What we mean by simulation of the market is a model that attempts to mimic the decision process of the principal agents—first buyers, wholesalers, retailers, etc.—and produces as a result estimates of the timing and magnitude of changes in prices, inventory holdings, sales, purchases, and so on at each level in the market.

7.4 Initial specification of the model

From the outset we had in our minds a relatively simple idea about market structure in which the agents at each level of the market bought, sold, and held product in response to price, supply, and high and low inventory signals at their own and other levels in the market (Figure 7.1). We began our work

[2] *Pounding* is the industry term for the inventorying of live lobster. The term is drawn from the tidal impoundments in which the inventorying takes place.

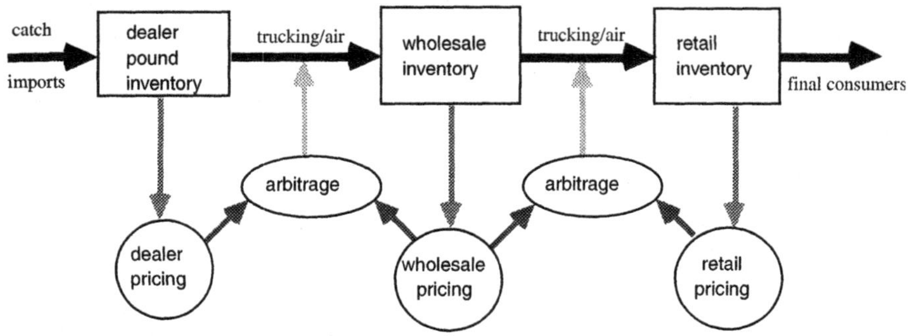

Figure 7.1 Preliminary structure.

in 1992 with extensive open-ended interviews of first buyers, wholesalers, and retailers.

The purpose of these interviews was to confirm our broad conception of market structure and dynamics and, especially, to determine the particular market signals agents acted upon. These initial interviews also pointed out the strong importance of context; that is, a price or other market signal may be very important under one set of circumstances, but at other times and circumstances the same signal of the same strength may be virtually irrelevant. For example, inventory levels are often important to the market but there are times in the late summer and early fall, when inventories are normally accumulated, that the level of inventories, per se, is relatively unimportant. In the late winter, however, when landings are close to non-existent, the same level of product inventory can be expected to have a strong influence on price expectations.

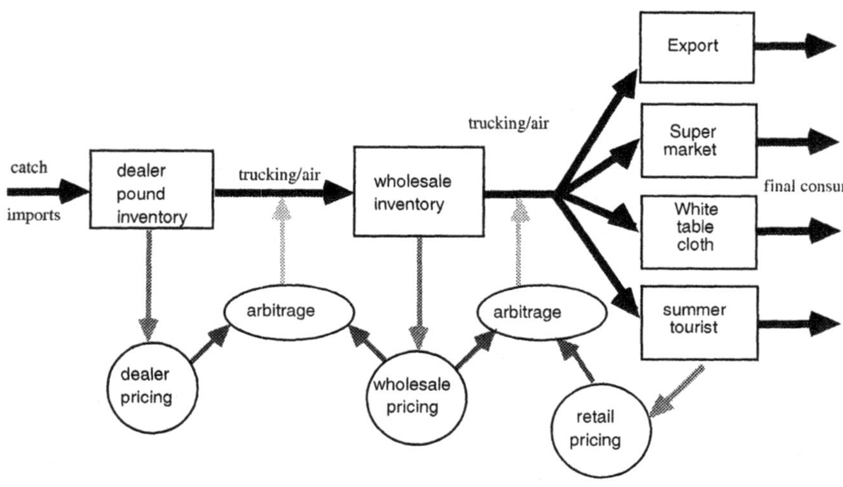

Figure 7.2 Approximate structure from interviews.

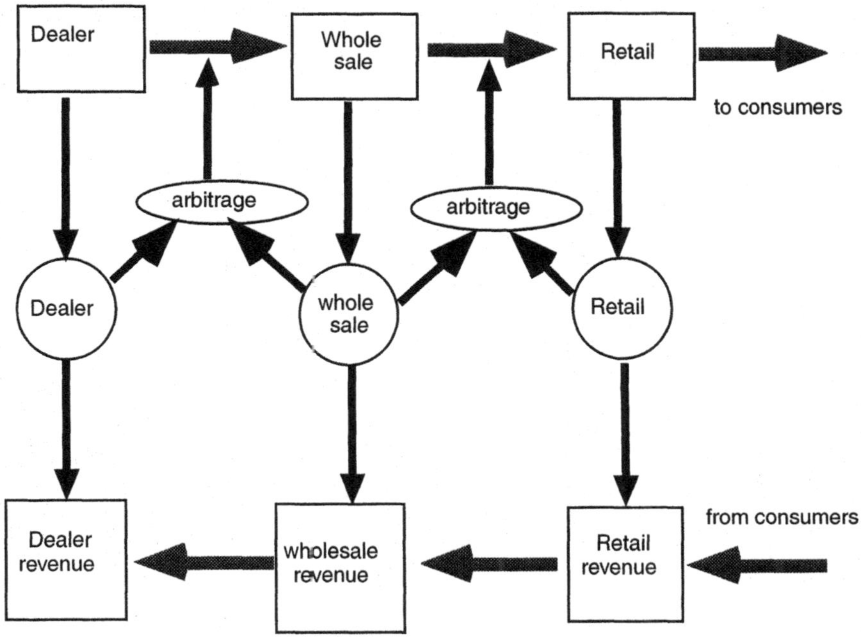

Figure 7.3 A three-level market with inventory, price and revenue flows.

On the basis of these interviews we constructed a fairly detailed flow diagram representing industry structure, product flow and decision points. This diagram identified nine different sectors within the market—dealers/first buyers and pounders at the first level of the market, wholesalers at the second level, and European exports, Asian exports, supermarkets, institutional services (Red Lobster, etc.), the summer New England tourist market and the year-round white tablecloth restaurant trade at the final retail level. (Figures 7.2 and 7.3) Industry informants further divided each of these retail markets into additional, finer categories on the basis of geographical or temporal variations. Similarly, the three levels in the market that we chose to categorize might have just as easily been set at four or more. A small (at that time) domestic processing sector producing frozen product was included in the dealer sector.

The next step in the process was to determine what quantitative and/or qualitative data existed or might be generated for the purposes of both specifying functional relationships and verifying the model. It was apparent from the outset that the descriptive detail about the market that we had accumulated far outran any existing data or the resources we had available to generate new data. We faced a situation where we had the simulation model equivalent of too many variables and not enough equations; even if the model were to predict well (which was highly unlikely) there was no way to verify the underlying structure—and that structure is what we were interested in.

Consequently, we set out with another set of open-ended interviews and a more formal survey for the purpose of determining the kinds and quality of data available and the level of disaggregation that might be supported in the model. There were two types of data we sought from these interviews: (1) qualitative data about the reaction of market agents to changes in market signals (e.g., the response of buyers to changes in the rate of landings and/or imports, i.e., supplies from Canada) and (2) data that might be used to verify the performance of the model (e.g., the timing of inventory peaks and troughs).

This second round of surveys quickly revealed consistent information about the general functional form of components in the model. For example, the flow of product between different levels of the market is basically an arbitrage process that is motivated by the profits to be made from price differences at the two levels of the market. Respondents uniformly agreed that widening price spreads provoked an accelerating rate of product flow. In short, the relationship between price spreads and product flow was seen as non-linear and increasing. (Figure 7.4) Importantly, what was not available from these interviews was any concrete sense of the quantitative aspects of the relationship. At first glance, this might seem to be a fatal omission, but as we show later, it is one that can be overcome more or less with the use of simulation.

One particular relationship derived from these surveys that is of strong interest concerns consumer response to changes in retail prices—sometimes known as the demand curve. At one point we had considered using monthly demand curves previously estimated by project members (Cheng and Townsend, 1993). However, we were surprised to find that informants at all levels of the market offered the very consistent information that when consumer prices fall below about $4.50 (all prices used here are 1981 real prices) sales volume accelerated rapidly. Above $4.50 sales were relatively steady

Figure 7.4 Generalized arbitrage function.

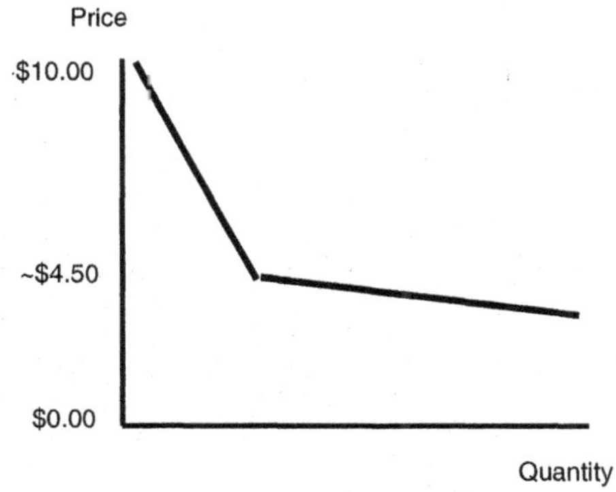

Figure 7.5 Generalized demand function.

but slowly declined until a price in the range of $9 to $10 was reached, at which time the market essentially shut down (Figure 7.5).

An interesting aspect of the survey response was that industry informants saw this schedule of demand prevailing over the entire market over the course of the year. In other words, although there was a typical seasonal pattern to demand (and prices) our informants argued that if prices fell below the $4.50 level at other than the usual times of the year, demand could be counted on to expand in the same fashion. Likewise, high prices at times of the year that normally saw low prices and high consumption would lead to a virtual 'shut down' of consumption. This demand schedule, like the other market response functions, could not be quantified on the basis of the surveys alone. Final determination of the quantitative attributes of the demand curve was estimated as described below through the use of the simulator.

7.5 *Qualitative criteria for checking the performance of the model*

These surveys also provided very valuable information for the purposes of verifying the performance of the model. 'Hard' published data on the market is relatively limited—only monthly data on landings and average monthly prices from 1980 to the present were available.[3] We were able to supplement this data with a variety of other, mostly qualitative, information which could be used to verify the performance of the model. For example, because of

[3] In the United States, data are collected independently by the various states and compiled by the National Marine Fisheries Service. 1980 is the earliest date for which consistent data for all the relevant states are available. In Canada, where the data series extends further back, data are collected by the Department of Fisheries and Oceans.

water temperature and market conditions, the long-term pounding of lobsters ends completely by the early part of July. Long-term inventories (pounded lobsters) tend to peak just before the Christmas rush, and so on. These 'check points' or criteria were developed so that when the simulator was run we would be able to see if it faithfully simulated these aspects of the market. All together we generated a list of some 12 qualitative checks of this sort. They are presented here:

1. Typical retail price behavior is characterized by:
   - stickiness
   - infrequent changes
   - adherence to traditional pricing points ($5.95, $6.95, etc.)
2. Retail demand appears to have three distinct ranges:
   - above approximately $9.00 demand appears to fall off rapidly with increases in price
   - in the range of about $4.50 to $9.00 demand appears to be relatively insensitive to changes in price
   - below $4.50 demand increases rapidly with decreases in price
3. Retailers prefer to keep small inventories relative to total turnover.
4. Retail price mark-up policies—retailers attempt to maintain constant gross margins but this may not be possible at both high and low prices.
5. Retailers are faced with much less active shopping around than sellers at other levels in the market. Consequently, retail prices deviate from one another much more than wholesale or exvessel prices.
6. Wholesalers/distributors tend to change their prices frequently with changes in exvessel prices in an attempt to maintain relatively constant margins.
7. Wholesale/distribution prices tend to be very competitive; the enforcing mechanism is the tendency for clients to maintain supply relationships with two or three wholesalers and to shop among them.
8. Wholesalers/distributors maintain much larger inventories than retailers (with improvements in transportation, these have been getting lower as well).
9. Dealers/pounds have the ability to hold large inventories in order to speculate against expected large swings in seasonal prices. Inventories (pound holdings) begin to accumulate in the early fall.
10. Pounds start to sell off in December and end in February or March, sometimes later.
11. A second inventory period occurs in the late spring, at which time Canadian and US spring catch is held in anticipation of the early part of the tourist season which occurs before the summer run of landings.
12. Dealers/pounds tend to face very competitive pricing. Prices change frequently (daily, sometimes hourly) in response to landings and other

factors. At this level of the market as well prices are kept competitive by clients shopping among a small number of regular suppliers.

In more traditional language, the demand curve is elastic above \$9.00, inelastic in the range between \$9.00 and \$4.50 and elastic again below \$4.50. (It is important to note that we obtained generally consistent descriptions of the general shape of the demand curve, including values on the price axis, but could not obtain the values on the quantity axis.)

7.6 Level of aggregation

In spite of the numerous checks that we were able to generate, one particular and, for purposes of disaggregation, critical set of data was not available. If one examines the detailed model diagram (Figure 7.1), it is immediately clear that a simulator constructed at this level of aggregation would need to simulate the relative flows of product to the various retail sectors outlined in the diagram. In the absence of this kind of information, any number of different allocations of product among the final markets might generate identical results elsewhere in the model, e.g., calculated exvessel prices. The second set of interviews and surveys made it apparent that even the most rudimentary quantification of the timing and volume of inventory flows to the various retail segments of the market would be very expensive to develop and, because of the proprietary nature of most of the information, maybe impossible.[4] Consequently, at this point we radically revised our design of the model. We reduced the detail to two market levels: a dealer/pound level at the first buyer level (hereafter referred to as the dealer sector), a single undifferentiated retail sector, and an arbitrage function that depended upon differences in prices at the two market levels (Figure 7.6).

The calculation logic of the model is this: supplies (both imports from Canada and U.S. landings) enter dealer/pound inventories. Changes in the rate at which supplies enter (which are treated as independent of the market), inventory levels and sales to retail (the rate at which supplies exit) act as signals that prompt dealers' prices to rise or fall. The rate of exit from dealer inventories depends on the difference between retail and dealer prices. Retailer prices are set on the basis of retail inventory levels and changes in inventories—both of which strongly reflect consumer demand. A change in retail prices causes a change in product flows from dealers which influences

[4] For many other industries, especially those that have developed inventory reporting systems, data of this sort are routinely available. If available, this kind of data would allow the analysis of the consequences of market promotion and other programs that target particular segments of the retail market and, consequently, when tied with a simulator might be a very valuable development tool.

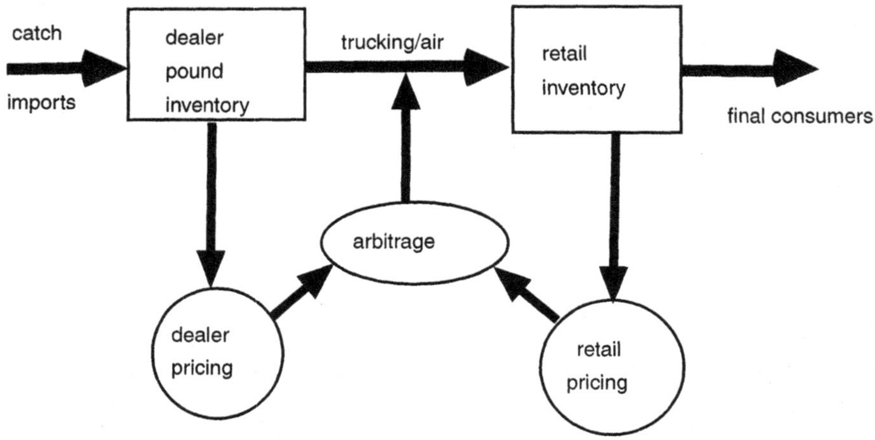

Figure 7.6 Final model structure.

dealer prices. In systems language, there are three interacting feedback loops—(1) dealer inventories influence dealer prices which influence arbitrage which in turn influences dealer inventories; (2) retail inventories influence retail prices which influences arbitrage which affects retail inventories; and (3) consumer demand influences retail inventories (through sales) which affects retail prices and so on.

7.7 Verifying the model structure

At this point in the process we had developed a model with a fairly concrete structure and a variety of criteria for verifying performance, but the specification of the principal functional relationships in the model was only given at a very general level. Our first step was to specify the quantitative attributes of these functions on the basis of *a priori* information about the market; in other words, we made educated guesses about the probable ranges over which the functional relationships operated. For example, we knew that from the individual dealer's perspective there was an acceptable level of inventories below which prices rose at an accelerating rate and above which the opposite occurred. Given a time horizon of a couple of weeks, we roughly estimated the acceptable level of inventories for the market as a whole to be in the range of 2 to 4 million pounds. Similar guesses were made with regard to the rate of price change in response to inventory levels. The same process was followed for all functional relationships.

Our next step was to rank the variables according to their temporal scale. The point of this ranking was to identify those variables that had the most pronounced effect upon the coarse grained annual behavior of the market as opposed to those whose principal effect was to influence fine grained short-term events.[5] Table 7.1 shows these rankings. We then began the truly Babylonian phase of the work. A time series of landings and prices composed of the same 'typical' year (1985–7 averaged) repeated 10 times was generated. These years were chosen in spite of the fact that we had fairly strong anecdotal evidence that a significant structural shift had taken place during 1990–91 when U.S.-landed lobsters began to be frozen in large numbers and, at the same time, there was a significant disruption in European and Asian exports because of the shortage of cargo aircraft during the Gulf War build-up. The point of this approach was to begin fitting the model to a set of data that was relatively free of strong exogenous shocks and also to test for the apparent structural shift.

Starting with the coarse grained variables—consumer demand and the arbitrage function—we used sensitivity analysis to search for quantitative attributes of these functions that would generate an annual inventory and exvessel price pattern that approximated those observed in the industry. When the most satisfactory set of values for these functions was located, a wide search over alternative values was conducted to check against having possibly located a sub-optimum.

We then followed a similar procedure with the next most coarse grained function—price response to dealer inventory levels—again searching for the most satisfactory set of values given our criteria for the performance of the model. When an apparent 'best set' of values was found, broad searches outside that range were conducted. Rather than continue down the ranked list of functions, it was necessary to return to the most coarse-grained functions—consumer demand and arbitrage—and repeat the search. Because all the functions in the model interact strongly—the strength of the interaction is roughly related to their proximity in the temporal ranking—any change in the dealer response to inventory levels can be expected to affect the nearby ranked functions of consumer demand and arbitrage.

This somewhat tedious procedure of slowly working down the ranking of functions and circling back to repeat previously conducted searches was followed until all the functions in the model were fully specified. To facilitate this search a number of calculations with graphical and tabular outputs were developed that provided quick indicators of the changing performance of the

[5] During the specification of the model it was very tempting to impose conditions that would force the model to meet the developed criteria. For example, if it is known that inventories reach a peak at a certain time of year, it is tempting to create a specification that imposes that result. Doing so, of course, will make the model perform better with regard to historical data for which this fact is known but at the same time completely disarm the model with regard to its ability to predict anomalous events that may lie outside the sample data.

Table 7.1 Temporal Ranking of Decision Variables and Their Effect on Simulation

Variable	Effect on model
• **Consumer demand**	
a. Horizontal shifts	• Horizontal shifts strongly affect dealer prices at all times of the year. Shifts to the right create upward pressure on estimated prices; shifts to the left, downward.
b. Vertical shifts	• Vertical shifts tend to shift sales between the upper leg of the demand curve and the lower. Consequently, upward shifts move more sales to the summer months, downward to the winter.
• **Arbitrage rate**	• Strongly affects the annual level and pattern of inventories, especially at the dealer level. Has equally strong impact on the spread between dealer and retail prices.
• **Dealer inventory level**	
a. Lower level	• Lower level tends to determine level of high prices in late winter.
b. Upper level	• Upper level tends to affect prices most in fall.
• **Changes in dealer supplies (landings and imports)**	• Tends to have strong effect in late fall (upward on prices) and strong downward in spring depending on when Canadian landings begin.
• **Changes in dealer inventory**	• Strong effects especially in the early spring and fall.
• **Changes in retail inventory level**	
a. Lower level	• Lower level generally weak short-term effects but felt throughout year. Tends to kick in at times of shortages and provokes jumps in both retail and dealer prices. Occurs most often in summer months when prices are low and "bursts" of retail sales occur.
b. Upper level	• Upper level affects retail prices strongly, but tends to have a weak effect upon dealer prices. Sporadic impact, no seasonal pattern.
• **Changes in retail inventory**	• Generally weak effects, no apparent time of year when this is especially important.

model. We did not keep track of the number of model runs required during this process but would estimate the number to be anywhere from 300 to 500.[6]

During this process we also maintained a continuing conversation with a number of informants in the industry. As mechanical as our description above may sound, it was anything but that. In effect, as we went through the process we began to find consistent problems with even the best quantitative specifications we could develop. Generally the problem was that we had missed a significant aspect of the functional relationship, for example, that during the month of June inventory levels even though very low, do not have much of an impact on the market because large landings can generally be expected within a few short weeks. Sometimes the nature of the problem became apparent during the search process and industry informants were used to confirm our suspicions; sometimes our informants provided completely new information that had not appeared before. This experience indicates a significant limitation to the model, and that is the ability to specify functional relationships as they may occur in unexpected circumstances lying outside the historical experience of our informants.

At the end of this process a model tuned very closely to our performance criteria was produced. Average absolute error in predicted exvessel prices was 3.9% while at the same time the qualitative criteria listed above were, by and large, met. Figure 7.7 shows the relatively tight fit of predicted and actual

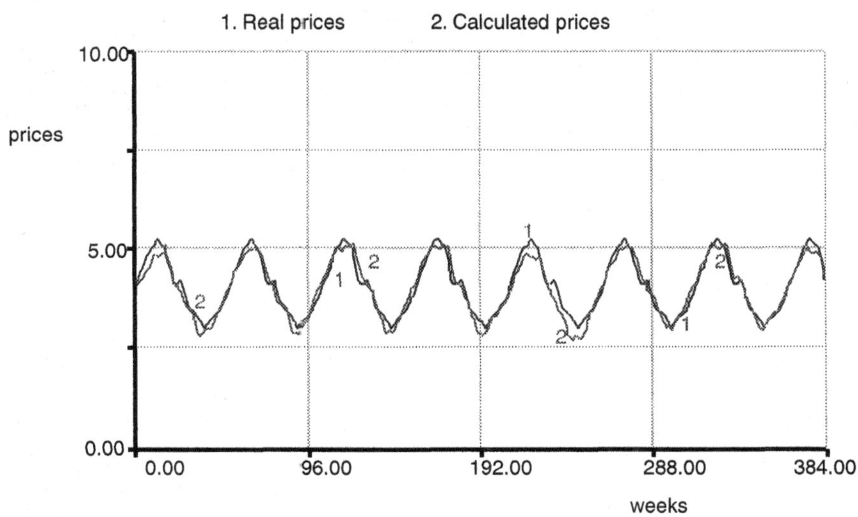

Figure 7.7 Real and calculated prices preliminary tuned model.

[6] Each 10-year, 520-period run requires less than 5 seconds to run. More significant, however, is the time required to analyze the graphical and tabular outputs and decide upon the next step in the search process—small or large increments, same or different direction.

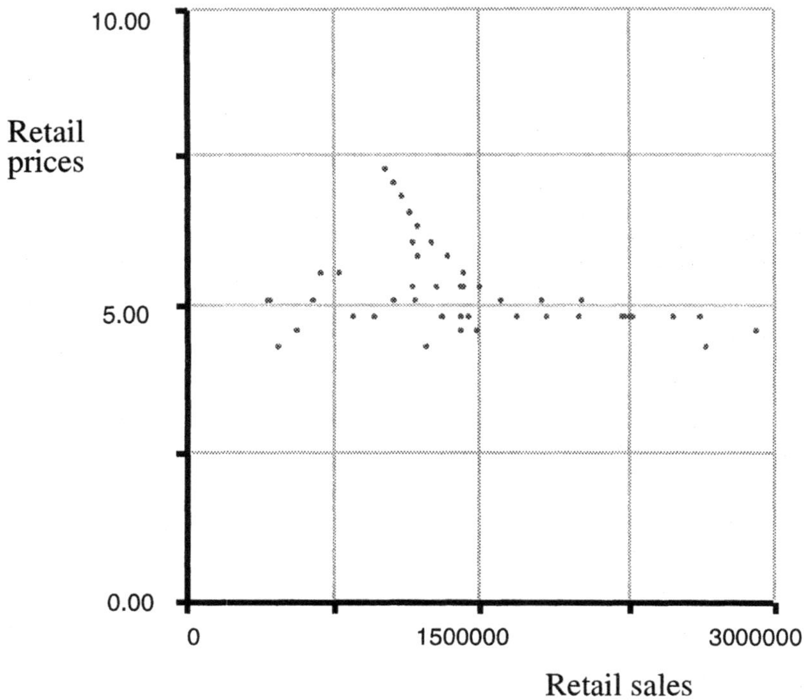

Figure 7.8 Estimated consumer demand curve.

prices. Figure 7.8 illustrates the retail price/quantity relationships generated by the model. "Observations" not lying on the demand curve were disequilibrium circumstances that provoked a price increase or decrease in the next period.[7]

7.8 *Testing the model with out-of-sample data*

This tuned model was then applied to data for the period 1986 to 1993. Figure 7.9 shows a time series graph of the results from that run.[8] The average absolute error in predicted exvessel prices was in the 10 to 11% range until 1989–90 after which it rises to nearly 20%. In short, the model does fairly well for 3 years beyond the sample and then falls apart. Even during the years when the absolute error is relatively low, there are relatively consistent ways

[7] The number of observations in Figure 7.8 is less than 520 because of the graphical resolution, i.e., many observations overlie one another at the resolution of the graph.

[8] The model is sensitive to initial values specified for inventories and prices. The circumstances produced by these misspecifications generally take a year or two to "work out." Consequently, every run of the model is initiated by three copies of the first year's data; i.e., 1986, 1986, 1986, 1987, 1988, . . . 1993.

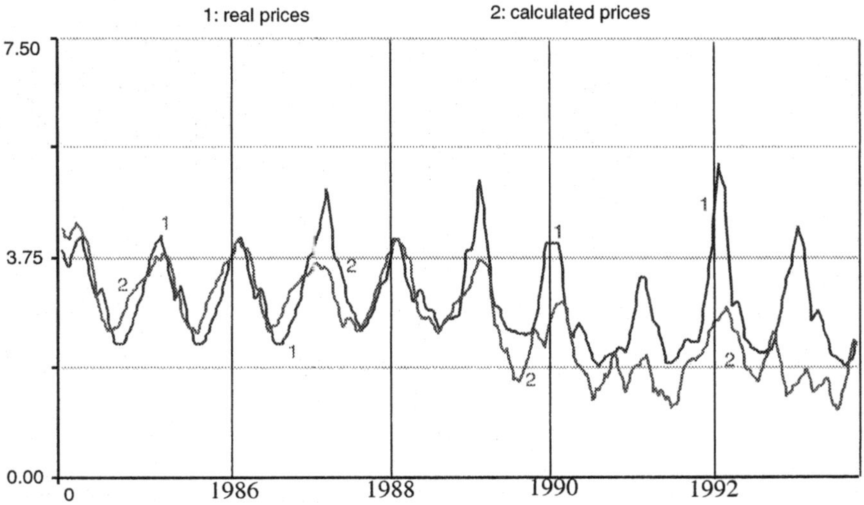

Figure 7.9 Model forecast and actual prices: first run of tuned model with out-of-sample data.

and times of the year when the error occurs. For example, in Figure 7.9 the model calculates the summer low point in prices reasonably well but every other fall generates a generally faster run up in prices. This tends to generate a lower rate of sales in the fall, less sell off from inventory, and (every other year) a tendency to miss the high late winter prices. In retrospect, these problems are attributable to either (1) the use of a "typical year" which suppresses the full range of variation in prices and supplies, and/or (2) inventory accumulations (in the model) that carry over from year to year and influence price changes.

7.9 Structural changes—the market for frozen product

As we expected, these initial runs show an apparent structural shift in the market that occurred in the 1989–90 period. At this time there was a very marked increase in landings over what had occurred throughout the 1980s—about 50%. The market responded to these increased landings by shifting a substantial amount of mostly Maine-landed product (upwards of 10 million pounds) into the frozen sector. Before this time freezing had been restricted to "undersized" lobsters landed in the Northumberland Straight area of the Gulf of St. Lawrence. At the same time a new market for this larger frozen product was opened.

In order to further explore our hypothesis about structural change we altered the model to include a small sector that reflected the decision process of processors of frozen product. As in our earlier work we interviewed industry informants. From them we learned that processors of frozen product worked on margins that made it possible for them to buy and operate when they expected exvessel prices to fall to the $2.25 to $2.50 range for 8 to 10 weeks (long enough to make it worthwhile to open the processing plants). Furthermore, processing capacity apparently was in the range of no more than 1 million pounds weekly. Figure 7.10 shows the results of altering the model to include a frozen sector.

This modification of the model produced absolute average errors in predicted exvessel prices in the range of about 11% with a very slight trend toward greater error in the later years. As in the earlier run (Figure 7.9), there is a consistent tendency for estimated prices to rise much earlier in the fall than actual and a (related) tendency to miss peak winter prices—a problem we trace back to the decision to tune the model on a "typical, averaged" year.

An examination of the demand curve estimated from this structurally revised version of the model (Figure 7.11) shows the "kink" in the demand curve to be somewhat lower than our survey information would lead us to expect. Additionally, the observations lying to the left of the "demand envelope" indicate periods of disequilibrium that are closely associated with those times when the model underestimates exvessel price. These disequilibrium situations are another perspective on the model's failure to track higher prices well. (The lines in Figure 7.11 connect successive price/quantity combinations calculated by the model. This tends to show the demand curve much more clearly and gives a better idea of the adjustment process in the model. For example, the lowest horizontal line represents a sudden drop in supplies; in the following month prices respond and there is a modest change in supplies, shown by the line moving to the northeast.)

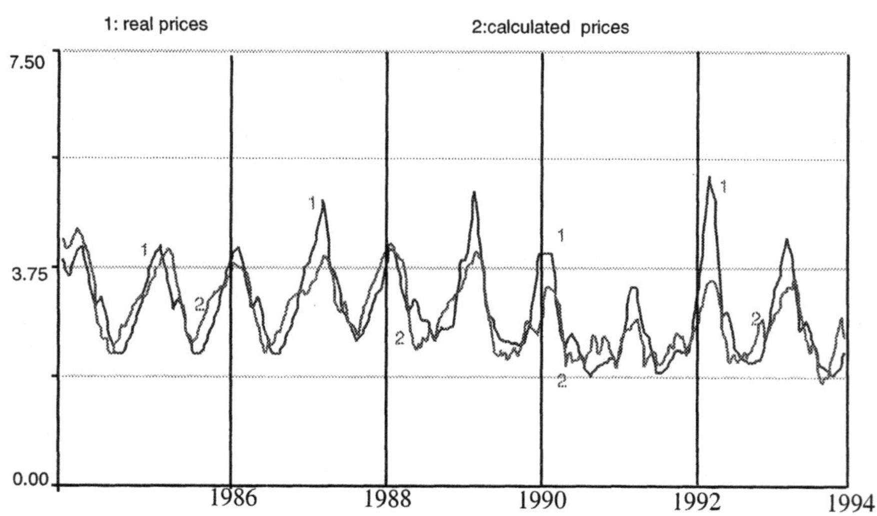

Figure 7.10 Model forecast and actual prices with market for frozen product.

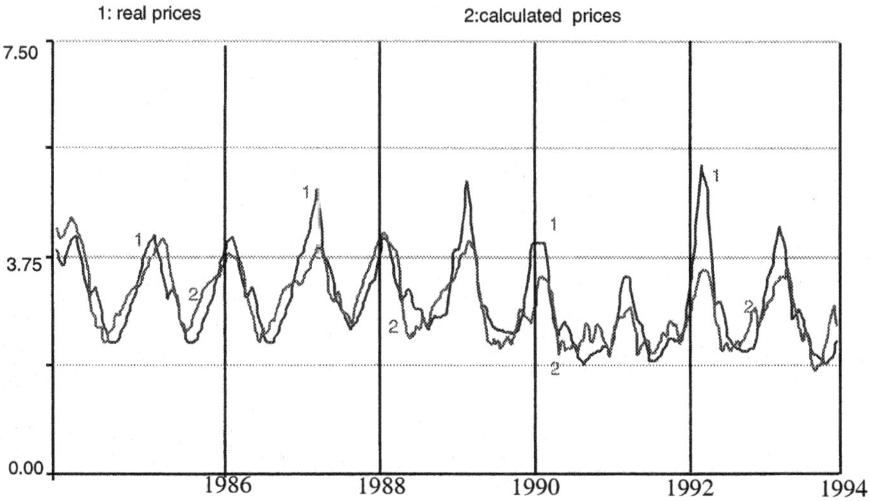

Figure 7.11 Estimated consumer demand curve with frozen product.

7.10 Summary

The model described here illustrates a way to use simulation modeling as an alternative to the economists' common problem of forecasting market behavior. Unlike statistical methods this approach requires fairly intimate knowledge of the market and a network of candid informants. (Alternatively, the requirements of constructing the model might be looked upon as a framework for learning about the operation and behavior of the market.) It allows the combined use of qualitative and quantitative data and, when necessary, permits the specification of very particular characteristics of the market, e.g., the effect of inventory levels in June or July as compared with the same in December or January. Like any model it has serious limitations due to its simplification of a complex environment. Figure 7.9, for example, shows the rapidity with which the quality of the model's forecasts deteriorates after the years over which the model was trained. And finally, although the forecasts of the model appear to be relatively good when compared with, say, typical statistical forecasts, there is no rigorous way to calculate the equivalent of confidence limits.

chapter eight

Modeling for scoping, research and management

Robert Costanza and Matthias Ruth

Contents

8.1 A three-step modeling process

The models presented to this point in the book have been designed to help understand a particular problem or process. This chapter describes a three step process for the use of models to support decisions about environmental investments and problems. Using models in this way greatly enhances the learning and communication process in a multi-stakeholder situation. The first stage of the process is to develop a high-generality, low-resolution **scoping and consensus building** model involving broad representation of stakeholder groups affected by the problem. STELLA and similar software make it feasible to involve a group of modeling novices in the construction of relatively complex models, with a few people competent in modeling acting as facilitators. Using STELLA, and projecting the computer screen onto the wall or sharing a model via the Internet, the process of model construction can be

1-5667-0389-1/01/$0.00+$.50

transparent to a group of diverse stakeholders. Participants can follow the model construction process and contribute their knowledge to the process.

After the basic model structure is developed, the program requires more detailed decisions about the functional connections between variables. This process is also transparent to the group, using well-designed dialogue boxes, and the potential for graphic and algebraic input. The models that result from this process are designed to capture as much "realism" as possible and to answer preliminary questions about system dynamics, especially its main areas of sensitivity and uncertainty, and thus to guide the research agenda in the following modeling stage.

The second stage **research** models are more detailed and realistic attempts to replicate the dynamics of the particular system of interest. This stage involves collecting large amounts of historical data for calibration and testing and conducting a detailed analysis of the areas of uncertainty in the model. It may involve traditional "experts" and is more concerned with analyzing the details of the historical development of a particular system with an eye toward developing specific scenarios or policy options in the next stage. It is still critical to maintain stakeholder involvement and interaction in this stage through the exchange of models and with regular workshops and meetings to discuss model progress and results.

While integrated models aimed at realism and precision are large, complex, and loaded with uncertainties of various kinds (Costanza et al., 1990; Groffman and Likens, 1994; Bockstael et al., 1995), our abilities to understand, communicate, and deal with these uncertainties are rapidly improving. It is also important to remember that while increasing the resolution and complexity of models increases the amount we can say about a system, it also limits how accurately we can say it. Model predictability tends to fall with increasing resolution due to compounding uncertainties (Costanza and Maxwell, 1993). What we are after are models that optimize their "effectiveness" (Costanza and Sklar, 1985) by choosing an intermediate resolution where the product of predictability and resolution (effectiveness) is maximized. As a consequence, resolution of the research models is medium to high, depending on the results of the scoping model.

The third stage of **management** models is focused on producing scenarios and management options in this context of adaptive feedback and monitoring, and based on the earlier scoping and research models. It is also necessary to place the modeling process within the larger framework of adaptive management (Holling, 1978) if management is to be effective. "Adaptive management" views regional development policy and management as "experiments," where interventions at several scales are made to achieve understanding and to identify and test policy options (Holling, 1978; Walters, 1986; Lee, 1993; Gunderson et al., 1995). This means that models and policies based on them are not taken as the ultimate answers, but rather as guiding an adaptive experimentation process with the regional system. Emphasis is placed on monitoring and feedback to check and improve models, rather than using models to obfuscate and defend a policy which does not correspond to reality. Continuing stakeholder involvement is essential in adaptive management.

Each of these stages in the modeling process has useful products, but the process is most beneficial and effective if followed in the order described. Too often we jump to the research or management stage of the process without first building adequate consensus about the nature of the problem and without involving the appropriate stakeholder groups. What we save on time and effort by jumping ahead is easily lost later on in attempts to forge agreement about results and generate compliance with the policies derived from the model.

8.2 Case studies

In this section we briefly describe a set of case studies that embody some or all of the characteristics of the three stage modeling process outlined above. The purpose of this section is to illustrate the wide range of environmental issues to which scoping and consensus building modeling has been applied, and to indicate the various degrees to which stakeholder involvement has been achieved in model development. We begin with case studies that solicited from stakeholders specific information to be included in the models and that shared throughout the modeling process the models with the contributors through a series of conversations, mailings, and presentations. We also present examples of cases in which workshop meetings for scoping and consensus building have been conducted in which a group of stakeholders convened to collectively develop models for scoping and consensus building purposes. Some of the models presented here have been followed up with more detailed research and management models.

8.2.1 U.S. iron and steel production

The iron and steel industry is the single largest energy consumer in the industrial sector of the U.S. economy and is characterized by large scale operations that require significant capital investment to change their structure and functioning. The high degree of interconnectedness among the various production stages often requires technological adjustments at one stage in response to change elsewhere in the industry. For example, many vertically integrated steel plants have been retiring their coke ovens, replacing them with imported coke and decreasing the production of pig iron, in which coke is used to reduce iron ores. The decline in pig iron production from blast furnaces is accompanied by a shift in raw steel production technologies away from those that use pig iron as their main input if overall raw steel output is to be maintained (Sparrow, 1983; Ross, 1987; Ruth, 1995). This typically means movement towards electric arc furnaces, whose main energy input is electricity. One effect of the changes in technologies at the various production stages is a significant change in the industry's energy use profile.

By-products such as coke oven gas and blast furnace gas have traditionally been used as energy sources in basic oxygen furnaces. The reduced production of pig iron leads to an increase in the fractions of energy purchased elsewhere rather than produced by the industry itself. The latter affects the industry's influence on its supply and cost of energy and has ramifications for its emissions profile.

The large investments that are required for the implementation of new technologies and the many interdependencies among the various production stages make it necessary for decision makers to anticipate long-term trends in demand of the industry's products and supply of raw materials and energy. By the same token, to move towards sustainable industry practices requires managers and policy decision makers to explore the implications for the industry's material and energy use profiles. They must explore a wide range of scenarios about changes in demand and the speed at which technologies can be adapted (Ruth and Harrington, 1997).

Using STELLA software, a model has been developed of iron ore mining, processing and raw steel production for the aggregate U.S. iron and steel industry with the goal to identify the industry's future likely profiles of material and energy use. The goal of the scoping and consensus building modeling of U.S. iron and steel industries was to capture the feedback among various production stages in the industry in terms of material and energy use. Particular attention was given to changes in material and energy flows in response to changes in input materials, technical change at the various production stages, and changes in demand for raw steel (Weston and Ruth, 1997). A series of informal, iterative interviews with industry experts, members of industry associations and consultants was carried out to arrive at a model structure that is sufficiently detailed to capture feedback responses and sufficiently simple to be easily communicated to non-expert modelers. Significant agreement was already present at the outset of the model development process on the system boundaries that define the respective production stages, and on the key material and energy types to be included in the model. Based on this consensus, the model captures mining, pig iron production and raw steel production, and modules for electricity generation and coke production (Figure 2.1 in Chapter 2).

To generate consensus on the specification of material and energy use at the various production stages and the feedback processes that occur among them, engineering information from various sources was used and supplemented with time series data derived from published sources. These quailifications provided a benchmark for model runs. To explore the industry's profiles of material and energy use under alternative assumptions, the model was set up to be run in an interactive modeling mode that enables decision makers to choose different parameter settings based on their understanding of the industry. Additionally, the model was designed to investigate the implications that various rates of change in demand for the industry's products and in technologies may have on material and energy use at individual production stages and by the industry as a whole.

The discussions with industry experts prior to setting up the model and running it indicated a prevailing assumption that even though crude ore reserves are finite, absolute amounts are large and ore grades sufficiently high to not pose a constraint on industry in the long run. Various model runs refuted this view of the industry. Even though there is no shortage of ore in the United States, the model shows that declines in ore grade lead to increases in total energy consumption per ton of raw steel output that is unlikely to be compensated for by improvements in technology—even in the presence of further increased recycling rates and only moderate demand increases. Valuable insight was generated with regard to changes in the industry's energy mix, technology mix and the time frames in which these changes are likely to occur.

Subsequently, the model has been significantly extended from the model designed for scoping and consensus building to include indirect energy requirements by the iron and steel industry, and direct and indirect carbon emissions (Ruth, 1995). Efforts are under way to work with managers in industry to guide investment decisions at the level of the firm and provide management support. Similar applications of dynamic modeling to industrial processes have been conducted for several other metals industries (Ruth, 1997), U.S. pulp and paper production (Ruth and Harrington, 1997), and U.S. container, flat and fiberglass production (Ruth and Dell'Anno, 1997).

8.2.2 Louisiana coastal wetlands

Applications of dynamic modeling for scoping and consensus building in industrial systems have concentrated on material and energy flows within these systems and between these systems and their environment. In contrast, the Louisiana coastal wetlands project traces the distribution of water and sediment through landscapes.

The changing historical sequence of the Mississippi River's main distributaries has resulted in sediment deposition that formed the current Mississippi deltaic plain marshes. This delta switching cycle lasts on average 1500 years and sets the historical context of this landscape. At present, the river is in the process of changing from the current channel to the much shorter Atchafalaya River. The U.S. Army Corps of Engineers maintains a control structure at Old River to control the percentage of Mississippi River flow going down the Atchafalaya. Since about 1950 this percentage has been set at approximately 30%. Atchafalaya River borne sediment first filled in open water areas in the upper Atchafalaya basin, and more recently has begun to build a delta in Atchafalaya Bay (Roberts et al., 1980; Van Heerden and Roberts, 1980a, 1980b). During the next few decades, a new delta is projected to form at the mouth of the river, and plant community succession will occur on the recently formed delta and in the existing marshes. At the same time, the overall Louisiana coastal zone is projected to have a net loss of approximately 100 km^2/yr due to sediment starvation and saltwater intrusion (Gagliano et al., 1981).

The leveeing of the Mississippi and Atchafalaya Rivers, along with the damming of distributaries, has virtually eliminated riverine sediment input to most Louisiana coastal marshes. This change has broken the deltaic cycle and greatly accelerated land loss. Only in the area of the Atchafalaya delta is sediment-laden water flowing into wetland areas and land gain occurring (Roberts et al., 1980; Van Heerden and Roberts, 1980a, 1980b).

Primary human activities that potentially contribute to wetland loss are flood control, canals, spoil banks, land reclamation, fluids withdrawal, and highway construction. There is evidence that canals and levees are an important factor in wetland loss in coastal Louisiana, but there is much disagreement about the magnitude of the indirect loss caused by them (Craig et al., 1979; Cleveland et al., 1981; Scaife et al., 1983; Deegan et al., 1984; Leibowitz, 1989). Natural channels are generally not deep enough for the needs of oil recovery, navigation, pipelines, and drainage, so a vast network of canals has been built. In the Deltaic Plain of Louisiana, canals and their associated spoil banks of dredged material currently comprise 8% of the total marsh area compared to 2% in 1955. The construction of canals leads to direct loss of marsh by dredging and spoil deposition and indirect loss by changing hydrology, sedimentation, and productivity. Canals are thought to lead to more rapid salinity intrusion, causing the death of freshwater vegetation. Canal spoil banks also limit water exchange with wetlands, thereby decreasing deposition of suspended sediments.

Proposed human activities can have a dramatic impact on the distribution of water and sediments from the Atchafalaya River, and consequently on the development of the Atchafalaya landscape. For example, the Corps of Engineers was considering extending a levee along the east bank of the Atchafalaya that would restrict water and sediment flow into the Terrebonne marshes.

This situation represented a unique opportunity both to study landscape dynamics and to build consensus about how the system works and how to manage it. The Atchafalaya landscape is changing rapidly enough to provide time-series observations that can be used to test basic hypotheses about how coastal landscapes develop. In addition to short-term observations, there is a long and detailed history of field and remotely sensed data available on the study area (Bahr et al., 1983; Costanza et al., 1983). Solutions to the land loss problem in Louisiana all have far-reaching implications. They depend on which combination of solutions are undertaken and when and where they are undertaken. Outside forces (i.e., rates of sea level rise) also influence the effectiveness of any proposed solution. In the past, suggested solutions have been evaluated independently of each other, in an *ad hoc* manner, and without adequate dialogue and consensus among affected parties.

In order to address this problem in a more comprehensive way, a project was started in 1986 to apply the three-stage modeling approach described above. The first stage of scoping and consensus building involved mainly representatives of the Corps of Engineers, the U.S. Fish and Wildlife Service, local landowners and environmentalists, and several disciplines within the academic community. This stage involved a series of workshops aimed at developing a "unit model" (using STELLA) of the basic processes occurring at any point in the landscape, and at coming to agreement about how to model the entire landscape in the later stages. This stage took about a year.

In the second (research) stage, an integrated spatial simulation modeling approach was developed (Costanza et al., 1988; 1990; Sklar et al., 1985; 1989); that replicated the unit model developed in stage 1 over the coastal landscape and added horizontal flows of water, nutrients, and sediments, along with successional algorithms to model changes in the distribution pattern of habitats on the landscape. Using this approach, the ability was demonstrated to simulate the past behavior of the system in a fairly realistic way (Costanza et al., 1990). This part of the process took about 3 years.

In the third (management) stage of the dynamic modeling process a range of projected future conditions was laid out as a function of various management alternatives and natural changes, both individually and in various combinations. The research and management model simulates both the dynamic and spatial behavior of the system, and it keeps track of several of the important landscape-level variables in the system, such as ecosystem type, water level and flow, sediment levels and sedimentation, subsidence, salinity, primary production, nutrient levels, and elevation.

The research and management model was called the Coastal Ecological Landscape Spatial Simulation (CELSS) model. It consists of 2479 1-km^2 spatial cells to simulate a rapidly changing section of the Louisiana coast and predict long-term (50 to 100 year) spatially articulated changes in this landscape as a function of various management alternatives and natural and human-influenced climate variations.

The model was run on a CRAY supercomputer from initial conditions in 1956 through 1978 and 1983 (years for which additional data were available for calibration and validation) and on to the year 2033 with a maximum of weekly time steps. It accounted for 89.6% of the spatial variation in the 1978 calibration data and 79% of the variation in the 1983 verification data. Various future and past scenarios were analyzed with the model, including the future impacts of various Atchafalaya River levee extension proposals, freshwater diversion plans, marsh damage mitigation plans, future global sea level rise, and the historical impacts of past human activities and past climate patterns.

The model results were used by the Corps of Engineers and the Fish and Wildlife Service in making decisions about these management options. Because they were involved directly as participants in the process through all three stages, the model results were much easier both to communicate and to implement. The participants also had a much more sophisticated understanding of the underlying assumptions, uncertainties, and limitations of the model, along with its strengths, and could use it effectively as a management tool.

8.2.3 South African fynbos ecosystems

While the Louisiana wetlands project concentrated on aspects of the physical landscape, another scoping and consensus building project was initiated to address issues of species diversity. The area of study is the Cape Floristic Region—one of the world's smallest and, for its size, richest floral kingdoms. This tiny area, occupying a mere 90,000 km^2, supports 8500 plant species of which 68% are endemic, with 193 endemic genera and 6 endemic families (Bond and Goldblatt, 1984). Because of the many threats to this region's spectacular flora, it has earned the distinction of being the world's "hottest" hotspot of biodiversity (Myers, 1990).

The predominant vegetation in the Cape Floristic Region is fynbos, a hard-leafed and fire-prone shrubland which grows on the highly infertile soils associated with the ancient, quartzitic mountains (mountain fynbos) and the wind-blown sands of the coastal margin (lowland fynbos) (Cowling, 1992). Owing to the prevalent climate of cool, wet winters, and warm, dry summers, fynbos is superficially similar to California chaparral and other Mediterranean climate shrublands of the world (Hobbs et al., 1995). Fynbos landscapes are extremely rich in plant species (the Cape Peninsula has 2554 species in 470 km^2) and the amount of narrow endemism ranks among the highest in the world (Cowling et al., 1992).

In order to adequately manage these ecosystems several questions had to be answered, including what services do these species-rich fynbos ecosystems provide and what is their value to society? A 2-week workshop was held at the University of Cape Town (UCT) with a group of faculty and students from different disciplines along with parks managers, business people, and environmentalists. The primary goal of the workshop was to produce a series of consensus-based research papers that critically assessed the practical and theoretical issues surrounding ecosystem valuation as well as assessing the value of services derived by local and regional communities from fynbos systems.

To achieve the goals, an "atelier" approach was used to form multidisciplinary, multicultural teams, breaking down traditional hierarchical approach to problem solving. Open space (Rao, 1994) techniques were used to identify critical questions and allow participants to form working groups to tackle those questions. Open space meetings are loosely organized affairs which give all participants an opportunity to raise issues and participate in finding solutions.

The working groups of this workshop met several times during the first week of the course and almost continuously during the second week. The groups convened together periodically to hear updates of group projects and to offer feedback to other groups. Some group members floated to other groups at times to offer specific knowledge or technical advice.

Despite some initial misgivings on the part of the group, the loose structure of the course was remarkably successful, and by the end of the 2 weeks, seven working groups had toiled feverishly to draft papers. One group focused on producing an initial scoping model of the fynbos. This modeling group produced perhaps the most applicable and well-developed product resulting from the workshop: a general dynamic model integrating ecological and economic processes in fynbos ecosystems (Higgins et al., 1997). The model was developed in STELLA and designed to assess potential values of ecosystem services given ecosystem controls, management options, and feedback within and between the ecosystem and human sectors. The model helps to address questions about how the ecosystem services provided by the fynbos ecosystem at both a local and international scale are influenced by alien invasion and management strategies. The model comprises five interactive sub-models, namely hydrological, fire, plant, management, and economic valuation. Parameter estimates for each sub-model were either derived from the published literature or established by workshop participants and consultants (they are described in detail in Higgins et al., 1997). The plant sub-model included both native and alien plants. Simulation provided a realistic description of alien plant invasions and their impacts on river flow and runoff.

This model drew in part on the findings of the other working groups, and incorporates a broad range of research by workshop participants. Benefits and costs of management scenarios are addressed by estimating values for harvested products, tourism, water yield, and biodiversity. Costs include direct management costs and indirect costs. The model shows that the ecosystem services derived from the Western Cape mountains are far more valuable when vegetated by fynbos than by alien trees (a result consistent with other studies in North America and the Canary Islands). The difference in water production alone was sufficient to favor spending significant amounts of money to maintain fynbos in mountain catchments.

The model is designed to be user-friendly and interactive, allowing the user to set such features as area of alien clearing, fire management strategy, levels of wildflower harvesting, and park visitation rates. The model should prove to be a valuable tool in demonstrating to decision makers the benefits of investing now in tackling the alien plant problem, since delays have serious cost implications. A research and management modeling exercise may ultimately follow from this initial phase.

8.2.4 Patuxent River watershed, Maryland

The three case studies described above concentrate on economic systems and aspects of the physical and biological environment with little emphasis on the feedback processes that relate economic and ecological systems. In contrast, the case study on the Patuxent River watershed, MD, explicitly addresses the combined ecological-economic system to scope environmental problems and build consensus. It is addressed in detail in Chapter 9.

8.3 Conclusions

The complexities that surround environmental investments and problems require that nonlinearities and spatial and temporal lags are reflected in models used for decision support. Dynamic modeling is designed to address these system features. It also lends itself as a method to scope environmental problems and build consensus and has been used in an array of case studies ranging from industrial systems to ecosystems to linked ecological-economic systems.

In each case study described above (and in the following chapter), the three–stage modeling process enabled us to provide a set of detailed conclusions regarding the management of the respective system. These conclusions were built on models that embodied the input and expert judgment of a broad range of stakeholders. The modeling process also offered unique insight into our ability to anticipate a system's dynamics in the light of nonlinearities, and spatial and temporal lags. Our ability to anticipate those dynamics on the basis of available data and knowledge, and to develop consensus about those dynamics, is an essential prerequisite for the successful management of complex ecological-economic systems. We anticipate that future modeling efforts will increasingly make use of the software tools and the three-step modeling process with stakeholder involvement described in this chapter.

chapter nine

Case study: Patuxent River watershed, Maryland

Robert Costanza, Alexey Voinov, Roelof Boumans, Thomas Maxwell, Ferdinando Villa, Lisa Wainger, and Helena Voinov

Contents

1-5667-0389-1/01/$0.00+$.50

In this chapter, we describe in detail the integrated modeling framework we developed to model the Patuxent River watershed. This framework was designed to address many of the problems and modeling concerns we raised earlier in the book (Chapter 1). With this model, we try to support the efforts to effectively manage the complex interactions between human and natural systems at the watershed scale. Specifically, we hope to support the "tributary strategy" of the Chesapeake Bay Program, which recognizes that each major tributary watershed in the bay has unique problems and characteristics and needs to be understood and managed appropriately. We hope this helps to develop a better predictive understanding of watershed-level ecosystems, including the processes and mechanisms that govern the interconnected dynamics of water, nutrients, toxins, and biotic components and their linkage to human factors affecting water and watersheds.

Our modeling approach evolved from work in coastal Louisiana (Costanza et al., 1990) and in the Everglades (Fitz et al., 1993). Current work is focused on the Patuxent River watershed in Maryland (Figure 9.1), one of the best studied tributaries of the Chesapeake Bay, and one that has often been used as a model of the entire bay system. In particular the modeling framework is aimed at addressing the following general and more specific questions. Some of these have been addressed and are described in this chapter. Others await future research.

1. What are the quantitative, spatially explicit and dynamic **linkages** between land use and terrestrial and aquatic ecosystem productivity and health? More specifically:
   - What are the projected impacts of various future land use development scenarios (including "business as usual,") on terrestrial and aquatic ecosystem productivity and health?
   - What is the relative impact of changes in area and changes in practices on agricultural and residential lands on their impacts on terrestrial and aquatic ecosystem productivity and health?
   - How effective is the 1000-foot critical area setback under various local site conditions?
2. What are the quantitative **effects** of various combinations of natural and anthropogenic stressors on ecosystems and how do these effects change with scale? More specifically:

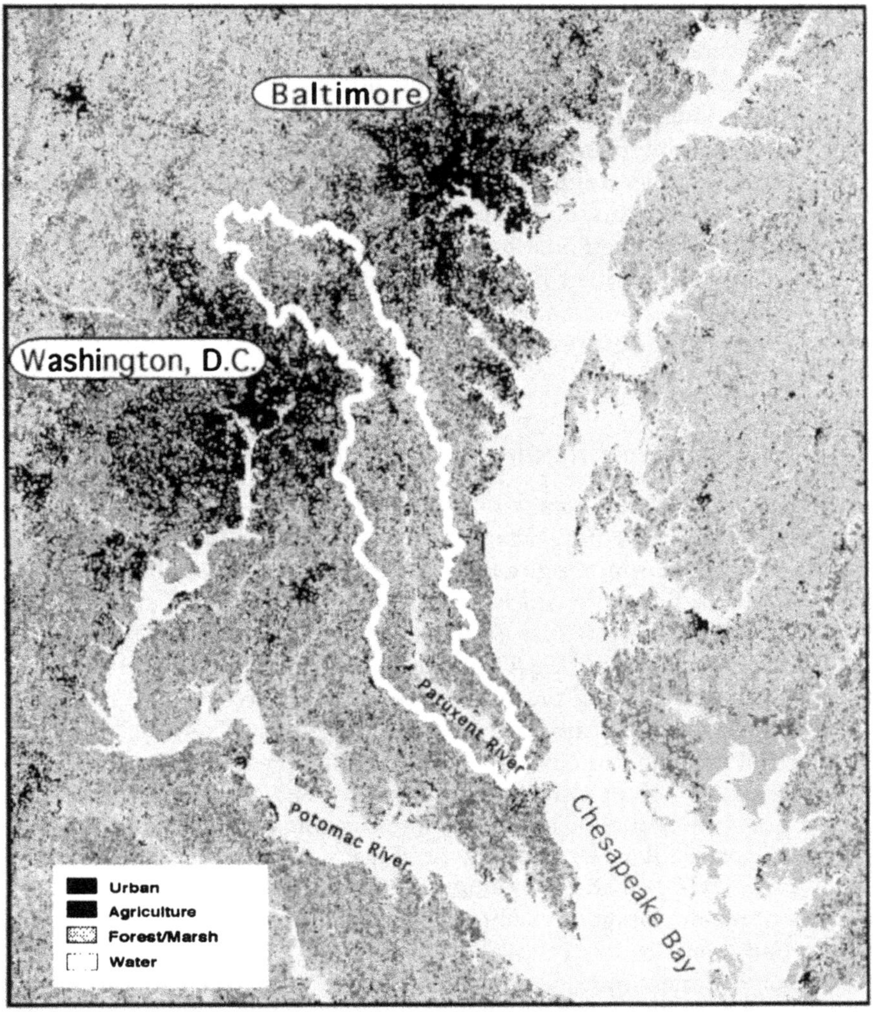

Figure 9.1 Location of the Patuxent watershed. Background map is based on NOAA C-CAP land cover data of Chesapeake Bay watershed 1988/89. Resolution is 30 m.

- What are the effects of nutrients, species additions and removals, and toxins, both alone and in various combinations on coastal ecosystems?
- Do these effects exhibit any regular, predictable changes with scale that can be used to extrapolate across scales?

3. What are useful ways to measure changes in the total **value** of the landscape including both marketed and non-marketed (natural system) components and how effective are alternative mitigation

approaches, management strategies, and policy options toward increasing this value? More specifically:

- Understanding how ecosystems function and how they are affected by human activity—for example, what determines human uses and human intervention into ecosystems, and how is this affected, among other things, by the ecosystem's characteristics and regulatory paradigms.
- Improving methods for ecosystem valuation—for example, what services provided by an ecosystem are of value to society, what methods should be used to measure this value, and in what terms should it be expressed.

9.1 *The Patuxent watershed: history and current setting*

Land use manipulation by humans began before European colonists first settled in the Chesapeake Bay. Native Americans burned the forest, removing underbrush and promoting growth of large trees (Mountford, 1997). However, the sediment record in the bay and its tributaries only begins to show the influences of land use change after European settlement in the 17th century (Brush and Davis, 1984; Cooper, 1995). It was not until massive land clearing and new farming practices were implemented in the 18th century that ecological effects became significant in the river and estuary.

Colonist farming practices, such as allowing livestock to range free, led to the local extinction of some herbaceous plants. Soil erosion increased dramatically as turning the soil before planting became a common practice in the 18th century, and tillage straight downhill was used to assist animal plowing (Mountford, 1997). As massive amounts of land were cleared, the estuary underwent major changes in sedimentation, eutrophication, and hydrologic regime. Sediment studies in a freshwater marsh (Jug Bay) in the Patuxent show that sedimentation rates went from 0.05 to 0.08 cm/year before European settlement to 0.50 cm/year on average in the mid-1800s, the time of maximum land clearance (Khan and Brush, 1994). Cores from the Khan and Brush study also show sharp increases in sedimentation rates in the mid-1960s through the mid-1970s and again in the early 1980s when rapid urbanization occurred in the upstream watershed. Rates have decreased somewhat since then, but current water depth at Jug Bay is only about 1 m according to their study.

Evidence for increasing eutrophication following European settlement is strong in the sediment record in the oligotrophic part of the Patuxent (Brush and Davis, 1984; Cooper, 1995). Biogenic silica, diatom diversity, and changes in relative abundance of certain species of diatoms all indicate eutrophication and some cores show peaks in eutrophication indicators at the time of initial land clearance for agriculture (around A.D. 1760). Concentrations of organic carbon, nitrogen and sulfur also show marked increases after this point (Cooper, 1995). Diatom species show a strong shift at the time that fertilizers were introduced in 1860 and sewage inputs to the river led to increases in total diatom abundance.

Submerged aquatic vegetation (SAV) decreased dramatically in the mid-1960s. This has been attributed to eutrophication and sedimentation (Kemp et al., 1983, 1985). Aerial photos from 1938 show large SAV beds in the Patuxent, some extending 300 m offshore (Mountford, 1997). The grasses have recently begun to reappear in significant amounts in some northern parts of the river, beginning in 1994 (Orth et al., 1995; VIMS Mapping Lab, 1997).

The changing land use has somewhat competing effects on the hydrologic regime. Removal of trees tends to reduce the evapotranspiration rates, leading to potentially higher freshwater inflows to the stream. Data from Baltimore Harbor in the Patapsco River just north of the Patuxent River (Figure 9.1), suggest salinity has decreased in the last 500 years, supporting this idea (Biggs, 1981). Yet, the effects of increased impervious surfaces can lead to decreased stream base flows by preventing groundwater recharge, although storm peak flows would be expected to increase. The exact nature of the hydrologic change through history is not well constrained and may be addressed by comparing runs of our model using historical and current land cover information (see further on).

Current land use practices in the basin are the subject of intense scrutiny, as planners and legislators search for ways to limit the impact of growth on natural resources. In 1994, about 50% of the watershed remained in natural vegetation (40% forest), and the developed land uses included 15% residential, 5% commercial/industrial, and 30% agriculture (Figure 9.2). The seven counties that encompass the watershed are experiencing some of the highest growth rates in the state. Forest loss from 1985 to 1990 was greater than 6000 acres in the majority of the counties, and agricultural land declined over the same period by at least 2000 to 4000 acres in most counties (MOP/MDE, 1993).

The effort to raise public consciousness about the environmental problems in the bay can be seen as progressing through three distinct stages (Costanza and Greer, 1995): (1) the era of shared experience and raised consciousness (1965 to 1976); (2) the era of intense scientific analysis with political backing (1977 to 1983); and (3) the era of implementation and monitoring (1983 to the present). It was not until the implementation era that the full extent of the bay's problems began to be realized and in particular the necessity to treat the entire bay watershed as a system. Nutrient reductions required changes in activities in the watershed. While point sources of nutrients, like industries and sewage treatment plants, were relatively easy to control, it soon became evident that non-point sources, like residences and agriculture, were responsible for a large part of the problem and were much more difficult to control. The growing population of the watershed itself came to be recognized as a primary cause of the bay's problems.

The State of Maryland now has significant experience with watershed-based approaches to water quality protection and restoration. This has come

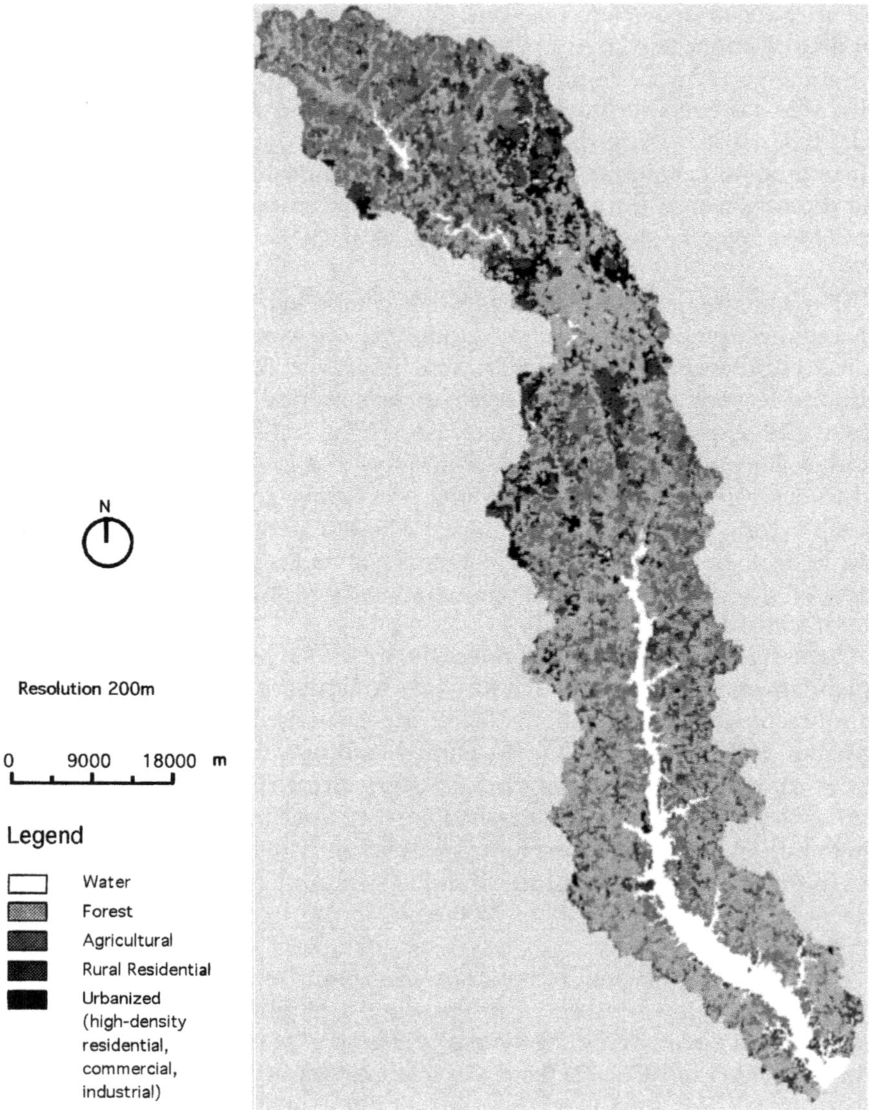

Figure 9.2 Patuxent Watershed Land Use Map for 1994. Based on the Maryland Counties Generalized Land Use/Land Cover maps acquired from Maryland Office of Planning.

about largely as a part of the multistate federal Chesapeake Bay Program and because the water quality of downstream tidal and estuarine waters has been recognized to be heavily influenced by upstream sources, particularly nonpoint sources. In addition, the nature of the Maryland portion of the bay watershed lends itself to delineation of discrete tributary watersheds, which include tidal rivers. A major focus of the program to restore the water quality of the bay involves a "tributary strategy" in which the sources of pollutants are estimated for each tributary watershed, fluxes are modeled, loadings are related to ecological conditions and living resources in the receiving subestuary, and goals are set for reduction of contaminants by generating sector (e.g., sewage treatment plants, agriculture, and dispersed residential) and location in the watershed. Thus the focus came to be on watersheds and individual tributaries to the bay. The Patuxent is one of the most important of these tributaries and a wealth of data has been collected within and near the river, as described below.

9.2 Methods

9.2.1 Watershed-based studies

Large drainage basins are composed of multiple smaller catchments. Each of these catchments contains a heterogeneous collection of land uses that vary in composition and spatial pattern (structure) and thus differ in functions such as nutrient retention. Two problems arise from this heterogeneity that present major challenges to both research and management. First, variation in structure and function inevitably prevents true replication in intensive field studies that attempt to relate landscape function to landscape structure. Second, variation among land uses within watersheds makes it difficult to directly extrapolate intensive studies to larger spatial scales. Even though drainage basins can be broken down hierarchically into smaller catchments based on topography, "scaling up" from intensive catchment studies is not a linear additive process because of differences among catchments and interactions between adjacent land uses. Management of water quality over large drainage basins must address both problems with innovative methods synthesizing data from intensive experimental studies on a few watersheds, then extrapolating important generalizations to large drainages using appropriate modeling techniques.

The most comprehensive approach to understanding nutrient flux from heterogeneous watersheds is through process modeling. Process models reflect understanding of the physical and ecological processes that either retain or release nutrients in watersheds. A process-based modeling approach to watershed-nutrient export linkages relies on (1) identification of ecological or anthropogenic processes important in making labile nutrients available for export, (2) simulation of hydrologic flow paths as potential "routes" of nutrient export, and (3) linking spatial patterns of labile nutrient concentrations with water flow paths. By focusing on processes, this approach gains potential generality of application at different spatial scales. Application of process models to any specific watershed depends on spatial representation of the landscape composition and topography, and promotes identification of specific nutrient export problems and selection of management actions to correct them. General process models require intensive data for development and independent data for testing. Thus, these models are best developed from intensive empirical and model calibration studies of contrasting sub-watersheds at several scales and tested by their ability to predict nutrient export from a variety of watersheds.

9.2.2 The Patuxent Landscape Model

The Patuxent Landscape Model (PLM) was designed to serve as a tool in a systematic analysis of the interactions among physical and biological dynamics of the watershed, conditioned on socioeconomic behavior in the region. A companion socioeconomic model of the region's land use dynamics was developed to link with the PLM and provide a means of capturing the feedback between ecological and economic systems. Because of the complex feedback and nonlinear dynamics of this watershed, a "systems" approach was necessary. A key part of this process was the development of an integrated, dynamic, spatially explicit simulation model.

In the ecological component of the model, the important processes that affect plant communities are simulated within the varying habitats distributed over the landscape. The principal dynamics modeled are (1) plant growth in response to available sunlight, temperature, nutrients, and water; (2) flow of water plus dissolved nutrients in three dimensions; as mediated by the (3) decomposition of dead organic material and formation of soil organics. Using this approach for incorporating process-based data at a reasonably high spatial, temporal, and complexity resolution within the entire watershed, the changing spatial patterns and processes can be analyzed within the context of altered management strategies, such as the use of agricultural best management practices (BMPs) (e.g., reduced fertilizer application and reduced tillage).

The modeled landscape is partitioned into a spatial grid (ranging in this application from 2352 to 58,905 square unit cells. The model is hierarchical in structure, incorporating an ecosystem-level "unit" model that is replicated in each of the unit cells representing the landscape (Figure 9.3). The unit model, referred to as the General Ecosystem Model or GEM (Fitz et al., 1996) itself is divided into submodels or modules representing functional divisions that simulate the important dynamics for each aspect of the system.

The current GEM unit model includes modules for hydrology, nutrient movement and cycling, terrestrial and estuarine primary productivity, and aggregated consumer dynamics. The hydrology sector of the unit model is a fundamental component for other modeled processes, simulating water flow vertically within the cell. Phosphorus and nitrogen are cycled through plant uptake and organic matter decomposition. The plant module includes growth response to various environmental constraints (including water and nutrient availability), changes in leaf canopy structure (influencing water transpiration), mortality, and other basic plant dynamics. Feedback among the biological, chemical, and physical model components are important structural attributes of the model. While the unit model simulates ecological

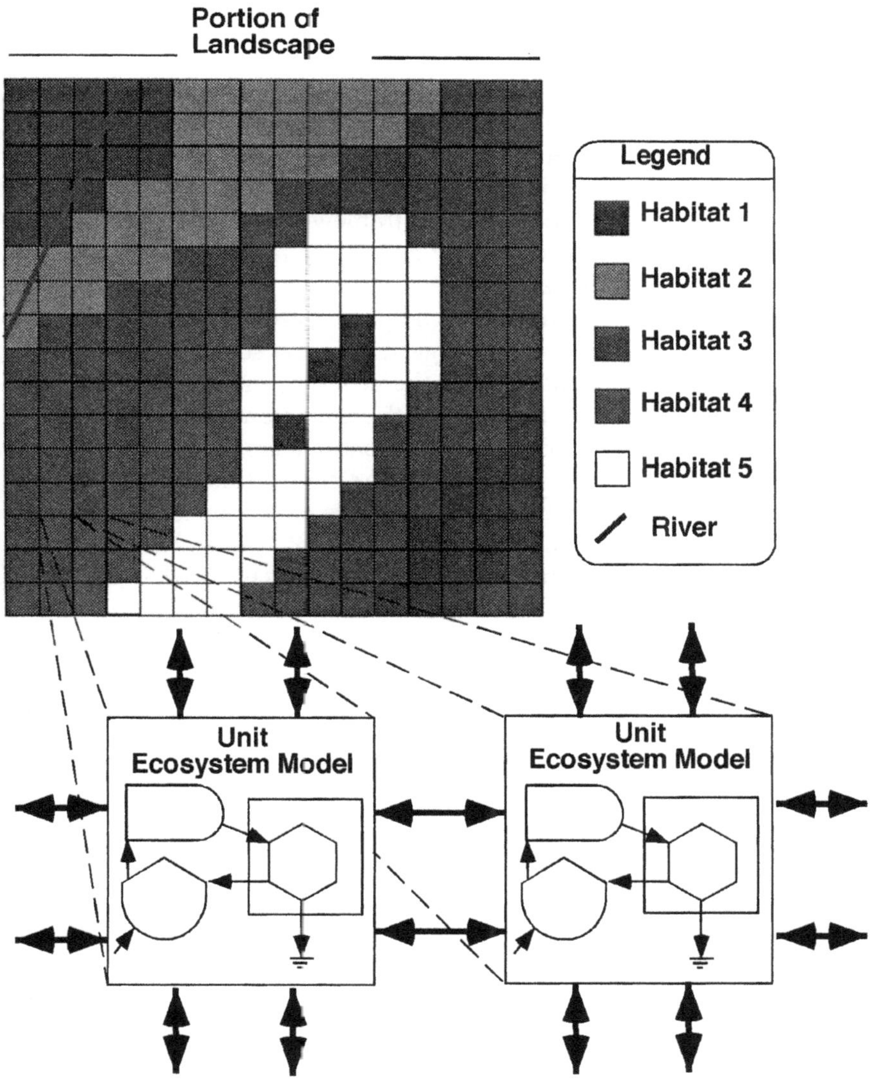

Figure 9.3 The cellular structure of the PLM. Each cell has a (variable) habitat type, which is used to parameterize the unit model for that cell. The unit model simulates ecosystem dynamics for that cell in the above-sediment and below-sediment subsystems. Nutrients and suspended materials in the surface water and saturated sediment water are fluxed between cells in the domain of the spatial model.

processes within a unit cell, horizontal fluxes link the cells together across the landscape to form the full PLM. Such fluxes are driven by cell-to-cell head differences of surface and ground water in saturated storage. Within this spatial context, the water fluxes between cells carry dissolved and suspended materials, determining water quality in the landscape.

The same general unit model structure runs in each cell. Individual cells are parameterized according to habitat type and georeferenced information for each cell is stored in GIS files. A habitat-dependent parameter database includes initial conditions, rate parameters, stoichiometric ratios, etc. The vegetation community type in the cells responds to changing hydrologic and nutrient regimes via successional switching algorithms (Costanza et al., 1990). Thus, when run within the spatial framework of the PLM, the landscape responds to changing hydrology and water quality as simulated by material flows between adjacent cells.

The ecological model is linked to a companion economic model that predicts the probability of land use change within the seven counties of the Patuxent watershed (Bockstael, 1996). The economic model allows human decisions to be modeled as a function of both economic and ecological spatial variables. Based on empirically estimated parameters, spatially heterogeneous probabilities of land conversion are predicted as functions of land values in residential and alternative uses, and costs of conversion. Land value predictions, themselves, are modeled as functions of local and regional characteristics. The model of land use conversion generates the relative likelihood of conversion of cells, and thus the spatial pattern of greatest development pressure. To predict the absolute amount of new residential development, the probabilistic land use conversion model is combined with models of regional growth pressure. Linking the ecological and economic models allows the effects of both direct land use change through human actions and indirect effects through ecological change to be evaluated, as well as the feedback between the two.

9.2.3 *Geographic and time-series data*

A variety of spatially and temporally disaggregated data is required to develop and calibrate the model. The database we have assembled is partially described in Table 9.1. The model database contains the data that drive the model forcing functions, parameterize equations, and provide calibration and verification data for adjusting model parameters and comparing model output to the real system. The database was developed from extensive data sets collected for the Patuxent watershed by various governmental agencies, academic institutions (Table 9.1), and research programs (Correll, 1983; Correll et al., 1992; Brush et al., 1980; Peterjohn and Correll, 1984; Lichtenberg and Shapiro, 1997). Existing data for the local region were supplemented with broader regional databases where appropriate.

Much of the available data is at a temporal or spatial resolution that is lower than we would like, so we sometimes employed data interpolation techniques to enhance the data. For example, maps of model driving forces such as precipitation are created as the model runs by interpolating time-series data from the set of seven meteorological stations distributed throughout the area. Another example is our use of elevation data (1:100,000, with

Table 9.1 PLM Data

Model inputs	Spatial resolution (no. of sites)	Temporal resolution/ time step	quality	Source
Time series				
Precipitation and Temp.	7 stations	50 yr/ daily	Good	EARTHINFO[1]
Wind speed, humidity	2 stations	5 yr/ daily	Good	EARTHINFO[1]
Habitat parameters				
Forest:				
Tree growth dynamics	Species level	1/yr	Good	NE TWIGS FVS[2,3]
Nutrient dynamics	E US/20 sites	1/yr	Good	Johnson and Lindberg[4]
Wetlands				
Nutrient retention rates	Pax R/6 sites	One time	Good	CEES, [5]JHU[6]
Stock values	225 locations	7 to 10 year	Good	CBP[7]
Population dynamics	Mesocosms	Biweekly	Good	MEER[8]
Agriculture				
BMP parameters:				
fertilizer applications	State	Annual	Good	UM-Ag. Extension[9,10]
Nutrient reduction/ retention rates	State/ county	Annual	Fair	CBP/UM-Ag. Extension[11]; MOP[12]
Population dynamics	1 point	4000 yr/ 1 d	Good	EPIC model data base[13]
Soil Interactions	0.1 km^2	Daily	Good	WEPP model database[14]
Urban				
% Impervious surface	Landuse/type	None	Good	MOP[38]; SCS[15]
Nutrients in runoff	Landuse/soil	event	Good	NURP US EPA[16]
Point source nitrogen	All NPDES	10 yr/ monthly	Good	MDE
Urban BMP efficiencies	Counties	Event	Fair	MOP[12]
GIS coverages				
Land use	200 m; 30 m	1984–94 / 5	Good	MOP[17]; NOAA; [18]EPA[19]
River network	200 m	None	Good	U.S. Census Bureau[20]
Soils	200 m	None	Good	MOP, STATSGO[32]
Elevation	3 arscec	None	Good	DEM USGS[22]
Watershed boundary	200 m	None	Good	Based on elevation

Estuarine bathymetry	200 m	None	Good	NOAA/NOS[23]
Roads and towns	Vector	None	Good	U.S. Census Bureau[20]
Groundwater (initial)	200 m	1985	Fair	USGS, [24]elev., river
Calibration Data				
Stream flow	13 stations	1979–95/daily	Good	USGS[25]
Surface water quality	13 stations	10 yr/bi-weekly	Good	CBP[26]; bi-ACB[27]
Groundwater levels	16 stations	5 yr/monthly	Good	USGS[24]
Groundwater quality	105 stations	1973–90, 1x/well	Fair	MDE[28]; USGS[29]
NDVI (Green Index)	1250 m	1993/monthly	Good	USGS[30]
Forest dynamics	187 sites	10 yr/10 yr	Fair	FIA[31]
Tree ring data	11 sites	175 yr/1 yr	Good	NOAA[32], IEE
Agricultural census data	State and county	50 yr/5 yr	Good	USDA[33]; USBCS[34,35]; DHMH[36]
Urban development	3 arcsec	1792–1992/13×	Good	BWRC[37]

1. EARTHINFO, I., 1993. *NCDC Summary of the Day, East 1993;* EARTHINFO, Inc.; Boulder, CO.
2. Bush, R.R., 1995. Northeastern TWIGS Variant of the Forest Vegetation Simulator. Forest Management Service Center.
3. USDA Forest Service, The Forest Vegetation Simulator (FVS) and Insect and Pathogen Models, http://162.79.41.7/fhtet/background.html.
4. Johnson, D.W. and Lindberg, S.E., 1992. *Atmospheric deposition and forest nutrient cycling: a synthesis of the integrated forest study.* Springer-Verlag, New York.
5. Zelenke, J.L. and Cornwell J.C., 1996. *Nutrient Retention in the Patuxent River Marshes.* Horn Point Environmental Laboratory, UMD, Cambridge, MD.
6. Kahn, H. and Brush, G.S., 1994. Nutrient and metal accumulation in a freshwater tidal marsh. *Estuaries,* 17(2): 345–360.
7. EPA Chesapeake Bay Program, *The Biological Data and Information Page,* http://www.epa.gov/docs/r3chespk/cbp_home/infobase/lr/lrsctop.htm.
8. Multiscale Experiment Ecosystem Research Center—MEERC, *Welcome to the MEERC World Wide Web (W3) Server,* http://kabir.cbl.cees.edu/Welcome.html.
9. Bandel, V.A. and Heger, E.A., 1994. *MASCAP*—Maryland's Agronomic Soil Capability Assessment Program, University of Maryland Cooperative Extension Service and Department of Agronomy in cooperation with USDA Soil Conservation Service, College Park, MD.
10. Coale, F., 1995. Fertilizer application rate and timing, Cooperative Extension Service, College Park, MD.
11. Steinhilber, P., 1995. FY 95 New Plan Fertilizer N, P_2O_5, and K_2O Reductions for Selected Crops, Nutrient Management Program.

12. Nonpoint Source Assessment and Accounting System, 1993. Maryland Office of Planning and Maryland Dept. of Environment.
13. Williams, J.R., Dyke, P.T., and Jones, C.A., 1983. EPIC: a model for assessing the effects of erosion on soil productivity, in *Analysis of Ecological Systems: State-of-the-Art in Ecological Modeling,* W.K. Laurenroth et al., Eds., Elsevier, Amsterdam, 553–572.
14. National Soil Erosion Laboratory, Water Erosion Prediction Project (WEPP). http://soils.ecn.purdue.edu/~wepphtml/wepp/wepptut/ahtml/avi/welcome.au.
15. Soil Conservation Service, 1975. Urban Hydrology for Small Watersheds, USDA Soil Conservation Service, Engineering Division.
16. U.S. Environmental Protection Agency, 1983. Results of the Nationwide Urban Runoff Program, Water Planning Division: Washington, D.C.
17. Maryland Office of Planning; Maryland State Data Center, htt://www.inform.umd.edu/UMS 1 State/MD_Resources/MSDC/.
18. USDC NOAA Coastal Service Center, *C-CAP.* Coastal Change Analysis Program. http://www.csc.noaa.gov/ccap/.
19. Vogelmann, J., 1996. *MRLC landuse coverages for 1992–93,* EROS Data Center.
20. U.S. Census Bureau, TIGER: The Coast to Coast Digital Map Database, http://www.census.gov/geo/www/tiger/.
21. NRCS USDA, State Soil GeographiC (STATSGO) Data Base, http://www.ncg.nrcs.usda.gob/statsgo.html.
22. USGS, USGS Digital Elevation Model Data, http://edcwww.cr.usgs.gov/glis/hyper/guide/usgs_dem.
23. NOAA/National Ocean Service, MapFinder. Preview Demonstration of August 1997 Offering, http://mapindex.nos.noaaa.gov/.
24. James, R.W.J., Horlein, J.F., Strain, B.F., and Smigaj, M.J., 1990. Water Resources Data. Maryland and Delaware. Water year Volume I. Atlantic Slope Basins, Delaware River through Patuxent River. U.S. Geological Survey, Towson, MD.
25. USGS, Maryland Surface-Water Data Retrieval, http://h2o.usgs.gov/swr/MD/.
26. EPA Chesapeake Bay Program, Water Quality Monitoring. Data Sets and Documentation, http://www.epa.gov/docs/r3chespk/cbp_home/inforbase/wqual/wqgate.htm.
27. Judd, M., 1996. *Citizen Monitoring Data,* Alliance for the Chesapeake Bay.
28. Wilde, F.D., 1994. Geochemistry and Factors Affecting Groundwater Quality at Three Storm-Water-Management sites in Maryland, Department of Natural Resources Maryland Geological Survey in cooperation with USGS, Maryland Department of the Environment and the Governor's Commission on Chesapeake Bay Initiatives.
29. McFarland, E.R., 1995. Relation of Land Use to Nitrogen Concentration in Ground Water in the Patuxent River Basin Maryland, USGS.
30. Jones, J., 1996. Normalized-Difference Vegetation Index. Research & Applications Group, USGS, Reston.
31. Hansen, M.H., Frieswyk, and T., Glover, J.F., and Kelly, J.F., *The Eastwide Forest Inventory and Analysis Data Base: Users Manual,* http://www.srsfia.usfs.masstate.edu/ewman.htm.
32. NOAA Paleoclimatology Program, Tree Ring Data, http://www.ngdc.noaa.gov.paleotreering.html.
33. Economic Research Service. U., *ERS Databases,* htt://www.econ.ag.gov/Prodsrvs/dataprod.htm.

34. Bureau of the Census, 1987 Census of Agriculture. Volume 1 Geographic Area Series. Part 20. Maryland, State and County Data. 1989, U.S. Department of Commerce, Washington, D.C.
35. Government Information Sharing Project: Oregon State University—Information service. Census of Agriculture: 1982, 1987, 1992, http://govinfo.kerr.orst.edu/ag-stateis.html.
36. Maryland Department of Health and Mental Hygiene, 1986. Description of the Patuxent Watershed Nonpoint Source Water Quality Monitoring and Modeling Program Data Base and Management System, MDE, Office of Environment Programs, Water Management Administration, Division of Modeling and Analysis.
37. Baltimore–Washington Regional Collaboratory, 200 Years of Spatial Data, UMBS, NASA, USGS, http://research.umbc.edu/~tbenja1/bwhp/main.html.
38. Maryland Office of Planning. 1973. Natural Soil Groups of Maryland Technical Report. MD Department of State Planning, December, No. 199.

1 m vertical resolution) combined with river network data (Maryland Office of Planning) to improve the watershed boundary and shoreline delineation.

Spatial (GIS) data include several types of data sets. One set of maps describes initial conditions, such as land cover, elevation, soil type, bathymetry, and groundwater elevation. Other spatial data developed from satellite images provide a time series of estimated ecological conditions, which are used for calibration purposes (e.g., Normalized-Difference Vegetation Index, NDVI, Jones, 1996). Watershed boundary, slope, aspect, and study area map layers were developed with the watershed basin analysis program in GRASS – Geographic Resources Analysis Support System (USACERL, 1993). Figure 9.4 shows the basic spatial coverages that have been employed in the PLM and some of the derived layers that were also essential for the hydrologic module. Spatial fluxes of surface water in watershed models are predominantly driven by the elevation gradient.

In addition to the meteorological time-series data that are used to map daily weather conditions, time-series data are used to provide other information at specific points in the landscape. For example, point-source discharges are used to introduce materials at specific points in the landscape. Hydrologic point time series data (stream flow, surface and groundwater quality) are used for calibration in the nontidal portions of the streams.

Specific rate constants are generally functions of spatial or habitat characteristics, such as soil or vegetation type. Habitat-dependent parameters include growth coefficients, uptake rates, and seasonal controls. About half of these data are specific to the Patuxent watershed with the remainder derived from a more general database or from literature sources.

9.2.4 Unit model

The General Ecosystem Model (GEM) that was developed for the Everglades Landscape Model (ELM) (Fitz et al., 1996) was modified for use within the

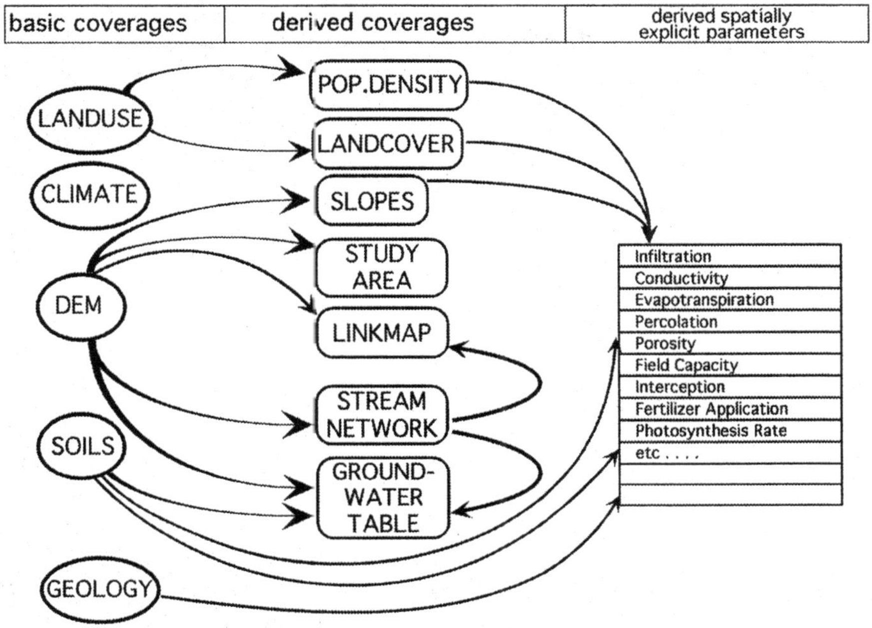

Figure 9.4 There are five basic PLM spatial coverages shown here. Additional maps are created during preprocessing and model initialization. Other spatial parameters and variables are calculated and updated during model runs.

framework of the PLM. The model was reformulated on a modular basis, with each module capable of being run and calibrated independently (Voinov et al., 1999b). The independent modules and the full unit model have also been run in the spatial implementation and rigorously tested at the full watershed scale. Some of the modules are described below.

Sensitivity analysis was used to gain insight about the model dynamics, showing the varying response of plant production to different nutrient requirements, with subsequent changes in the soil water nutrient concentrations and total water head. Changes in the plant canopy structure resulted in differences in transpiration, and consequently water levels and plant production.

9.2.4.1 Hydrology module

The hydrologic module simulates vertical water fluxes for a locality that is assumed to have spatially homogeneous characteristics. The module takes into account a variety of hydrologic functions controlled by physical and biotic processes including the following:

- Vertical water movement between surface, unsaturated and saturated storage from percolation, aquifer-stream interactions, and evapotranspiration.

- Spatial climatic forcing based on rainfall, temperature, humidity, and wind condition data.
- Transpiration fluxes dependent on plant growth, vegetation type, and relative humidity.

The traditional scheme of vertical water movement (Novotny and Olem, 1994), also implemented in GEM, assumes that water is fluxed along the following pathway: rainfall → surface water → water in the unsaturated layer → water in the saturated zone. In each of the stages, some portions of the water are diverted due to physical (evaporation, runoff) and biological (transpiration) processes, but in the vertical dimension the flow is controlled by the exchange between these four major phases.

At a daily time step, the model does not attempt to mimic the behavior of shorter term events such as the wetting front during an infiltration event. During a rapid rainfall event surface water may accumulate in pools and litterfall, but in a catchment such as the Patuxent, over the period of a day, most of this water will either infiltrate, evaporate, or be removed by horizontal runoff. Infiltration rates based on soil type within the Patuxent watershed range from 0.1–10 m/day, generally accommodating all but the most intense rainfall events in vegetated areas. The intensity of rainfall events can strongly influence runoff generation, but climatic data are rarely available for shorter than daily time steps. Also, if the model is intended to be run over large areas for many years, the diel rainfall data become inappropriate and difficult to project for scenario runs. Therefore, a certain amount of temporal detail must be forfeited to facilitate regional model implementation. With these limitations in mind, we have simplified the unit hydrologic model (Figure 9.5) in the following ways.

- We assume that rainfall infiltrates immediately to the unsaturated layer and only accumulates as surface water if the unsaturated layer becomes saturated or if the daily infiltration rate is exceeded. Ice and snow may still accumulate.
- Surface water may be present in cells as rivers, creeks and ponds. Surface water is removed by horizontal runoff or evaporation.
- Within the 1-day time step, surface water flux will also account for the shallow subsurface fluxes that rapidly bring the water distributed over the landscape into the micro channels and eventually to the river. Thus, the surface water transport takes into account the shallow subsurface flow that may occur during rainfall, allowing the model to account for the significantly different nutrient transport capabilities between shallow and deep subsurface flow.

9.2.4.2 Nutrient module

Nutrient enrichment is arguably the most important environmental degradation factor within the Patuxent watershed, especially in the estuary. The

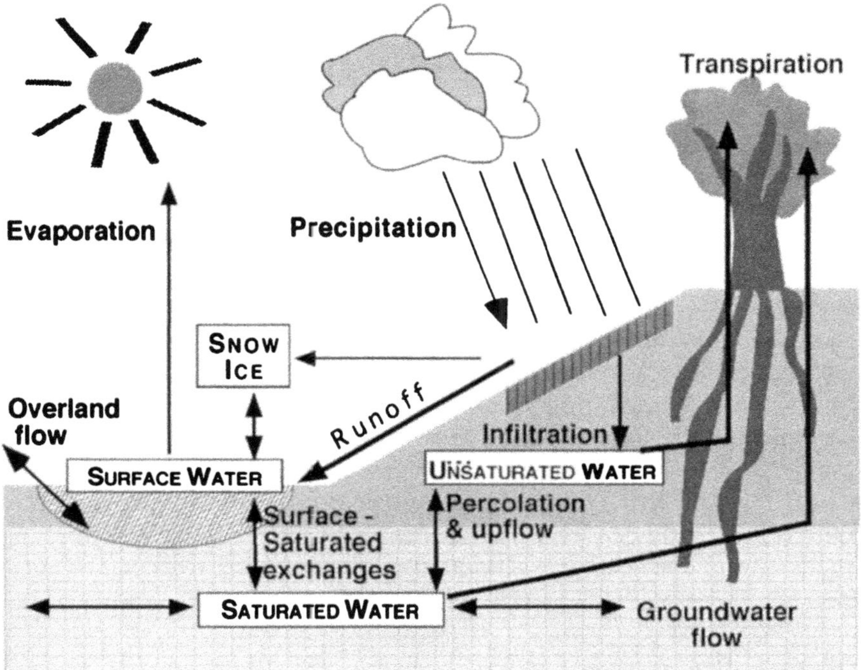

Figure 9.5 The unit hydrological model. The state variables are *Surface Water, Snow/Ice, Unsaturated Water,* and *Saturated Water.* The major processes are precipitation, evapotranspiration, infiltration, and percolation.

PLM process-based algorithms are designed to track the quantities of labile forms of phosphorus and nitrogen through the landscape during a simulation. Only phosphorus and nitrogen are tracked as they are the most likely nutrients to limit plant growth rates. Nutrient inputs of anthropogenic origin respond to socioeconomic forces and the effects of these inputs can be observed in plant production within the various habitats.

The original GEM nutrient sector was adapted to better match the aggregated hydrologic module. Various nitrogen forms, NO_2^-, NO_3^-, and NH_4^+ were aggregated into one variable representing all forms of nitrogen that are directly available for plant uptake. Available inorganic phosphorus is simulated as orthophosphate. The distinction appears in conceptualizing nutrients on the surface, since in the PLM they are no longer associated with surface water and therefore need not be in the dissolved form. Instead, since most of the time most of the cells have no surface water, the model variables represent the dry deposition of nitrogen or phosphorus on the surface. Over dry periods they continue to accumulate nutrients from incoming fluxes from air deposition or mineralization of organic material. When rainfall occurs, a certain proportion of the accumulated nutrients become dissolved and made available for horizontal fluxing and infiltration.

Further modification of the nutrient dynamics was required to accommodate the aggregation of surface and shallow subsurface flows in the hydrologic sector. In the PLM, a fraction of the nitrogen and phosphorus stored in the upper soil layer is made available for fast horizontal fluxing along with nutrients on the land surface. We have assumed this layer to be 10 cm thick, following a similar formalization in the CNS model (Haith et al., 1984), where this upper soil layer was also assumed to be exposed to direct surface runoff.

In addition to atmospheric deposition, nutrient sources include sewage discharges, fertilizer applications, and discharges from septic systems. At present, nutrient storage in biomass is estimated from empirical data on 12 US east coast mixed forest plots (Johnson and Lindberg, 1992). Dissolved inorganic nutrients are removed from the system through the growth of biomass and released through mineralization of soil or suspended organics. Nutrient uptake and release are modeled as proportional to total net primary productivity and decay at rates estimated from elemental carbon ratios measured in similar ecosystems.

9.2.4.3 Plant module

In the plant module we simulate the growth of higher vegetation. This will be macrophytes in an aquatic environment, trees in forests, crops in agricultural habitats, and grasses and shrubs in grasslands. The plant biomass (kg/m^2) is assumed to consist of photosynthesizing (PH) and non-photosynthesizing (NPH) components. In addition, we distinguish between above ground and the below-ground biomass and production.

Another state variable is employed to track "biological time" in the module. Biological time is the sum over the life span of the plant of the amount of time during which daily average temperatures exceed a certain value (5°C in our case). These are the temperatures that are most suitable for the physiological development of the plant. Therefore the total time during which such temperatures occur is a good indicator of the plant life stage and may be used to trigger certain processes such as sprouting, appearance of reproductive organs, and others.

The plant sector models conversion of inorganic carbon and nutrients into specific forms of biomass, and provides linkages to the hydrology through evapotranspiration. Maximum uptake rates are derived from empirical data relative to seasonal temperatures. During the simulation, maximum uptake rates are limited by light, nutrient concentrations, and water availability. The resultant uptake quantity is derived by multiplying the resultant uptake rate by the total photosynthetic potential (total leaf area).

The maximum attainable leaf area is habitat specific, and derived empirically as a proportion of the total biomass. Biomass and nutrients are accounted for in both above- and below-ground material. Once the optimum leaf to biomass ratio is reached during plant growth, excess biomass is routed to the non-photosynthetic component. The increase of total biomass allows additional increases in photosynthetic biomass. The non-photosynthetic biomass feeds back into the photosynthetic biomass by means of early spring sap flows in deciduous trees and through seed germination.

Plant mortality provides an input to detrital matter. Depending on the lignin content, we distinguish between stable and labile detritus. Stable detritus is shredded and becomes soil organic material or becomes labile detritus. Both soil organics and labile detritus are further decomposed to produce nutrients available for plant uptake.

9.2.4.4 *Human dominated systems*

Nutrients generated by human activities are of particular concern in the Patuxent River Watershed because as a Chesapeake Bay tributary it has been targeted for a 40% reduction in nutrient input levels. Agricultural and urban land use management are a focus of efforts to reduce non-point source nutrient inputs to water bodies throughout the Chesapeake Bay watershed. Research indicates that relatively simple, low-cost methods of farm nutrient management could have a significant impact on surface water and groundwater quality in the United States (Correll, 1983; Shabman, 1988; Crum et al., 1990; Lynch and Corbett, 1990; Magette et al., 1990) but the effects at the landscape scale of changing farming practices are not known.

The PLM can directly analyze changing land management practices for their effect on nutrients, water budgets, and plant ecosystems. Agricultural and urban practices that influence nutrient loading and water flow rates are expressed in the variables that influence infiltration rates and sinks and sources of nutrients. Agriculture is modeled using the plant sector of the model with special variables to account for fertilizer application and harvest times. To overcome the lack of spatially explicit data on farming practices, we used rates that are specific to soil type and county and rely on general information from those working with farmers and assessing nutrient practices in the basin. Since nutrient application rates are related to crop yields and since yield estimates vary by soil, we were able to estimate application rates for each parcel of farmland identified in the land use map, by linking recommended nutrient application rates (Bandel and Heger, 1994) to maps of soil and county (Figure 9.6). We plan to further refine the estimates of fertilizer application rates using the available data on county-level farming practices such as proportions of crops grown, rotation patterns, tillage types, and manure use. Currently we assume a crop rotation that is standard for this area. It includes corn, winter wheat, and soybeans over a 2 year time period.

Urbanization effects are largely handled through land cover-dependent parameters for the hydrology and nutrient modules of the GEM unit model. Available information on urban nutrient sources is linked to existing GIS layers such as roads and unsewered residential development. Point source discharge information, available in the form of time series data is input to the

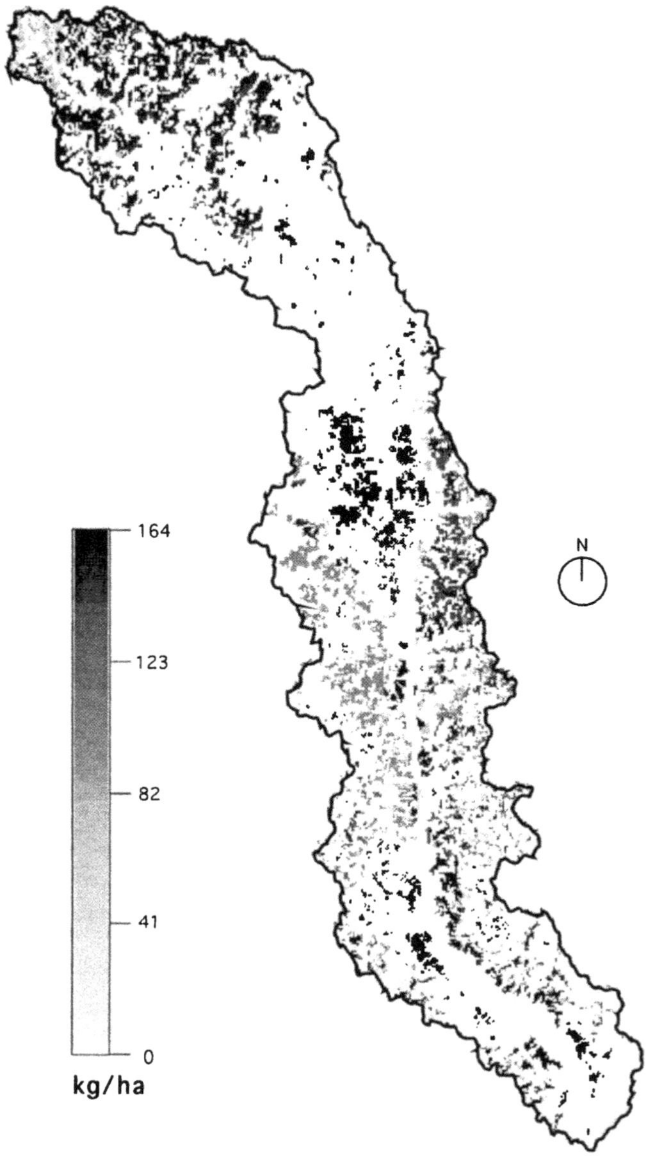

Figure 9.6 Map of fertilizer applications, generated based on the soil types, and expected yields for various counties. Sources : Maryland Natural Soil Groups information, soil surveys for Patuxent Watershed Counties; MD agronomical soil capability assessment program (defining yield expectations); plant nutrient recommendations based on soil tests and yield goals.

appropriate cells in the PLM. Information on distribution and effectiveness of stormwater management is used to parameterize residential land cover cells. The spatial model accounts for the effects of flow path on nutrient movement.

As nutrient-containing water moves from urban to undeveloped cells, the nutrients in surface water can be removed. Parameters on the type of soil, slope, and vegetation present in a cell determine whether that cell acts as a net source or sink for nutrients.

9.2.5 Spatial model

The spatial model combines the dynamics of the unit model (which are calculated at each time step for each cell in the landscape) and adds the spatial fluxes which control the movement of water and materials between cells. Each cell generates stock and flow values, which provide input to or accept output from the spatial flux equations.

In the spatial implementation, a major hypothesis that we are testing is that overland and channel flow can be modeled similarly. Traditionally, in most models of overland flow the surface water is moved according to two separate algorithms: one for the two-dimensional flux across the landscape and another for the one-dimensional channel flow. This approach is used in some of the classic spatial hydrologic models such as ANSWERS (Beasley and Huggins, 1980) or SHE (Abbott et al., 1986). However, considering the spatial and temporal scale of the Patuxent model, as well as its overall complexity, we use a simplified water balance algorithm for both types of flow.

Given the cell size of the model (200 m or 1 km), we may assume that in every cell there will be a stream or depression present where surface water can accumulate. Therefore, it makes sense to consider the whole area as a linked network of channels, where each cell contains a channel reach that discharges into a single adjacent channel reach. The channel network is generated from a link map, which connects each cell with its one downstream neighbor out of the eight possible nearest neighbors (Figure 9.7).

After the water head in each raster cell is modified by the vertical fluxes controlled in the GEM unit model, the surface water and its dissolved or suspended components move between cells based on one of two algorithms being tested. In the simplified algorithm a certain fraction of water is taken out of a cell and added to a cell downstream. This operation is either iterated several (10 to 20) times a day, effectively generating a smaller time step to allow fast riverflow, or the location of the recipient cell is calculated based on the amount of head in the donor cell after which in one time step the full amount of water is moved over several cells downstream along the flow path determined by the link map. The number of iterations or the length of the flow path needed for the hydrologic module is calibrated so that the water flow rates match gauge data.

Another algorithm checks that water movement stops when the water heads in two adjacent cells equilibrate. We examine the flow between two adjacent cells as flow in an open channel and use the *slope-area method* (Boyer, 1964), which is a kinematic wave approximation of St. Venant's momentum equation. The flux (m^3/day) in this case is described by the empirical

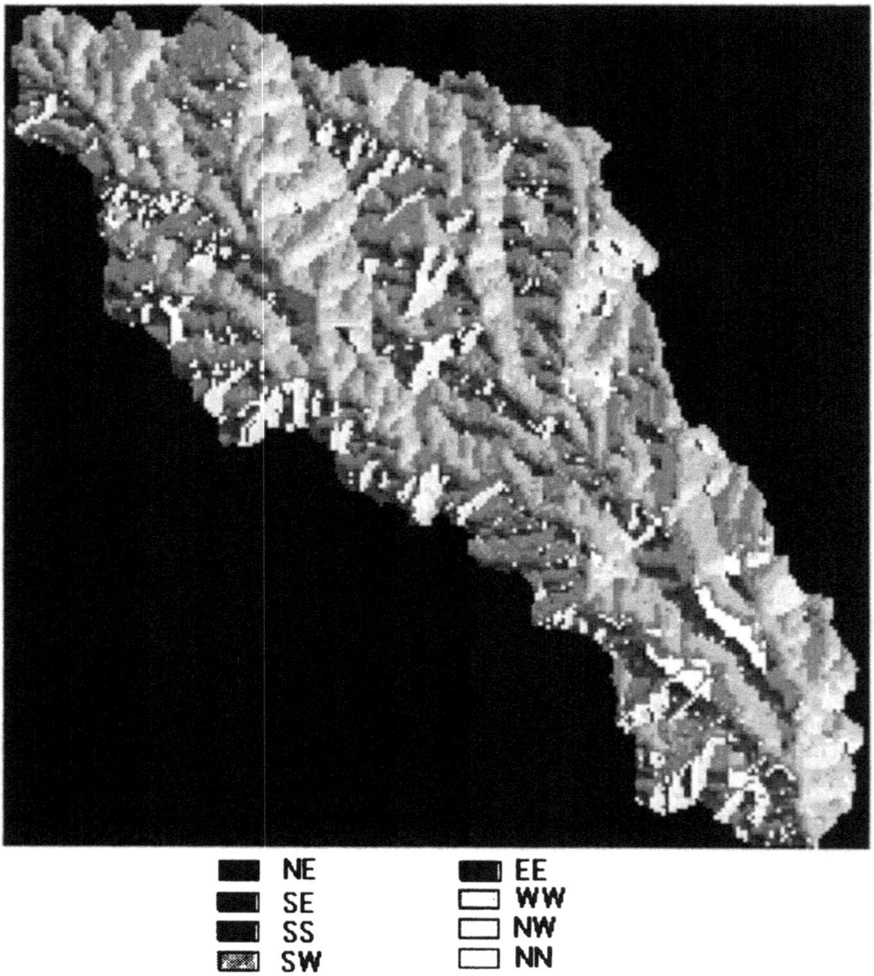

Figure 9.7 Link map. Each cell is marked by a number that indicates the direction of the next cell along the path of surface water flow: 2 = NN, 3 = NW, 4 = WW, 5 = SW, 6 = SS, 7 = SE, 8 = EE, 9 = NE.

Manning's equation for overland flow. The equation is further modified to ensure that there is no flux after the two cells equilibrate and then the flux rate is accelerated using the multicell dispersion algorithm discussed in Voinov et al. (1998). While the first algorithm works well for the piedmont area with significant elevation gradients, the second one is more appropriate for the coastal plain region where there are significant areas of low relief and tidal forces that permit counterflows.

For the saturated water, a modified Darcy equation was employed. For each cell, the flux was determined as a function of saturated conductivity and water head difference between the current cell and the average head of the cell and its eight neighbors. We assume one vertically homogeneous aquifer interacts with the surface water.

9.2.6 Economic land use conversion model

Spatially explicit data on property sales and many of the economic and ecological characteristics of areas in spatial proximity to these sales were available for the seven counties encompassing the Patuxent watershed. This allowed human decisions with regard to land use change to be empirically modeled in a spatially disaggregated way. The result is a model that estimates probabilities of land conversion from forest or agriculture to different densities of residential use within each spatial cell in the seven-county area of the Patuxent basin.

The first step of this process was to estimate, statistically, models that explain the value of land parcels in different uses. Prior work has been used to approximate values in agriculture and forest, but new models of the value of land in residential use were also developed (Bockstael and Bell, 1997; Bockstael, 1996; Bockstael et al., 1995; Geoghegan et al., 1997). This was made possible by an extensive GIS database that includes geo-coded records of all parcels in the tax assessment databases of the seven counties. The tax assessment database includes historical data on actual transactions (selling prices) together with characteristics and location of parcels. In addition, an extensive spatial database of land use, zoning, and other natural and human-imposed characteristics that might influence values in residential and alternative uses, as well as conversion costs, has been assembled.

Using sale transactions data, the assessed value of any structures was subtracted from the selling price to get the residual value of the land in residential use. This land value was used as the dependent variable and spatial variation in land prices were explained by an extensive array of features of the location: distance to employment centers, access to public infrastructure (roads, recreational facilities, shopping centers, sewer and water services), and proximity to desirable (e.g., waterfront) and undesirable (e.g., waste dumps) land uses to name a few. Also included were some less obvious explanatory variables that describe the nature of the surrounding land uses around a parcel. The estimation techniques used were maximum likelihood and generalized method of moments, the latter being an approach that allows for treatment of the obvious spatial autocorrelation in the model (Bell and Bockstael, 2000).

The ability to spatially locate transactions and account for locational characteristics, explicitly, has provided an improved technique to test assumptions about what affects residential land values. The model demonstrates the importance of scale (e.g., in lot size or development density) considerations and the nonlinearities associated with distance effects. For example, proximity to some features of the landscape, such as major highways, are positive amenities up to a point, but become disamenities when too close. This first-stage modeling exercise also provided a means of creating spatial maps of the estimated value of undeveloped land were it to be put into residential use, given the existing set of zoning ordinances, public utilities provision, highway network, etc. These predictions were then used in the second modeling stage.

The second stage involved estimating "qualitative-dependent variable" models (i.e., discrete choice models) of historical land use conversion decisions. In this stage, historical decisions of whether to convert a parcel in agricultural or forest use to residential use are modeled as functions of the value in original use, predicted value in residential use (derived from the first stage model), and proxies for the relative costs of conversion. The purpose of this model was to determine what factors affect land use conversion and to estimate parameters of those conversion functions.

Once the parameters of the two stages of the model were estimated, the model was used to generate the relative likelihood of conversion of different parcels in the landscape. Thus a spatial pattern of relative development pressure was obtained as a function of the characteristics of the parcels and their locations. Since the explanatory variables used to predict the values in residential and alternative uses and the costs of conversion are all functions of ecological features, human infrastructure, and government policies, the effects of changes in any of these variables can be simulated. Total growth pressure in the region was then used to estimate the spatial patterns of new residential development into the future.

9.2.7 *Linked ecological and economic models*

The linked ecological economic model can then be used to test several types of hypotheses about how future ecosystem behavior and urban development will be impacted by various policies, especially ones regarding zoning regulations and population increases. The economic side of the model generates probable land development patterns in response to regulations and population increases and the ecological model estimates the effects of these land use patterns on hydrology, forests, agriculture, and the estuary. Some of these responses can then be fed back into the economic model. With this linked model, the effectiveness of regulatory tools can be addressed by developing scenarios. The predictive model of human decisions permits testing of whether a proposed planning option such as extending sewer service to a targeted growth area will (1) be effective in luring development to that area and away from areas where growth is less desirable and (2) what the ecological impacts will be, such as nutrient levels in waterbodies.

Since land is traded in markets, the effect of changing ecological conditions on the private values of the land in different uses can also be approximated. However, many other ecosystem services may be valuable, but may not be appropriated by the private owners of land and thus will not show up in market values. These ecosystem services include the indirect ways that ecosystems support recreation and aesthetics, wildlife habitat, water supply, storm abatement, nutrient cycling, and human health, among others (Costanza et al., 1997).

9.2.8 Landscape pattern analysis

In addition to the indicators being developed from monitoring data, we are also using spatial landscape indices (Turner, 1989; O'Neill, 1988) to link simulation model output to ecosystem processes, which are not modeled spatially in the ecological model. Landscape-level analysis of land use data has shown promise in addressing how land use pattern may influence population abundance, diversity, and resilience (Geoghegan et al., 1997). Work in the Patuxent region has shown a correlation between bird abundance and species diversity with land use characteristics such as fragmentation (Flather and Sauer, 1996). Others have shown how source population distance and natural corridors can influence recovery of both plants and animals following a catastrophic event (Detenbeck et al., 1992; Gustafson and Gardner, 1996; Hawkins et al., 1988). We are calculating several spatial pattern indices using coastal plain watersheds and creating empirical models, which link population characteristics to spatial patterns. These empirical models can be applied to the Patuxent watershed and, more importantly, to changes in the watershed as predicted by the PLM.

9.3 Results

9.3.1 Calibration and testing

Much of the time involved in developing spatial process-based models is devoted to calibration and testing of the model behavior against known historical or other data (Costanza et al., 1990). To adequately test model behavior and yet reduce computational time, we performed the calibration and testing at several time and space scales, and for the unit model independently of the full spatial model. These tests and their results are briefly described below.

9.3.2 Model performance index

Very complex simulation models such as the GEM and PLM are now becoming computationally feasible, but the difficulty of finding parameter combinations and reproducing known data (calibration) increases sharply with complexity (Oreskes et al., 1994; Rykiel, 1996). In particular, it is widely recognized that models exhibiting moderate to high degrees of nonlinearity, when calibrated to real data, tend to show multiple, equally acceptable optimal configurations of parameters (Beven, 1993; Villa and Costanza, forthcoming). The model's response to parameter change is likewise a highly complex multivariate function whose characterization is difficult. Complex models often include a large number of variables, but calibration data exist on only a fraction of these and the data often do not match the time and space resolution of the model.

We developed a model performance index (MPI; Villa, 1997) to address the above problems. The MPI framework allows one to develop an error function that can handle the full range of variables and data quality that usually confront complex models. It employs a multicriteria approach, which allows user weighing of the model variables to reflect their degree of importance and also weighting the data to reflect its quality. It can deal with both quantitative and semiquantitative information about the expected behavior of the state variables (like the pattern of temporal autocorrelation, boundaries, steady state, etc.).

In the MPI, each variable is given a standardized partial "score" expressing the degree to which the calibration goals have been achieved. The single-score MPI is a weighted average of the partial scores. Techniques to search the parameter space of a complex model include Monte Carlo simulation, univariant deterministic search (e.g., hill-climbing), stochastic search algorithms (e.g., simulated annealing), and stochastic multivariant techniques (e.g., genetic algorithms).

We have also developed a model evaluation toolkit, a suite of parameter search tools using the MPI formalism. The toolkit assists model calibration and testing and allows one to quantify a model's complexity and predictability, as well as determine information about all aspects of a model's response to parameter change. Model calibration using this approach is a multistep, hierarchical process. General sets of criteria incorporate the general dynamic attributes (variable boundaries, seasonal dynamics, steady state) required for basic agreement with a naturally occurring situation. Such criteria are used for the initial characterization of the feasible areas of the parameter space. More specific calibration criteria, reflecting dynamic peculiarities at different sites, require the state variables to match the actual data recorded at different sites. The resulting multivariate surfaces (MPI value as a function of parameter values) are compared to identify areas of the parameter space that best explain differences and similarities among sites.

9.3.3 Unit model calibration

The GEM model has been previously calibrated for the Everglades National Park and water conservation areas (Fitz et al., 1995), which host a plant community dominated by annual species (e.g., sawgrass). In contrast, the dominant Patuxent plant community is forest. We have carried out a systematic investigation of the behavior of the GEM across its parameter space using our MPI, described above. The calibration presented here simulates a 10-year time period using a constant weather regime for 1986 and for each subsequent year. Field monitoring at 12 forested sites located within the eastern United States (Johnson and Lindberg, 1992) provided mean flux rates and organic matter nutrient contents for input and calibration. Input parameters for driving the hydrological functions are mean values from the STATSGO (Table 9.1) database for the Patuxent watershed. Biomass and species composition were derived through the Forest Inventory and Analysis Database (FIA) (Table 9.1). The data the forest association used were oak–hickory with 0.6% coniferous trees. External input of nutrients was through precipitation, while dissolved concentrations were allowed to leave the system with the groundwater flow. The consumer sector was made inactive in anticipation of stronger supporting data currently being developed.

The calibration was run for three different stages in forest development. At the first or young stage the forest biomass is set at 10% of the maximum attainable biomass, which is based on the 75th percentile value for oak–hickory in the FIA. The second stage (intermediate) is set at 50% of the maximum biomass, while the third stage (old) is set at 90% of the maximum biomass. Ten year averages of inorganic phosphate concentrations (PO^4), dissolved inorganic nitrate concentrations (DIN), net primary production (NPP) (Table 9.2), detrital matter and nonliving soil organic matter (NLOM) (Figure 9.8) are compared to similar values available through the FIA database for the Patuxent watershed, or literature on temperate forests.

As expected, the older forest showed the largest amounts of detrital and soil organic matter. The soil organic matter values in the old and intermediate forest exceed the mean + 1 SD (Figure 9.8) because of the lack of consumer appropriation of plant biomass, which is instead routed to the organic matter pool through the detrital pathway. In contrast to the intermediate or young forests, old forest was able to sustain the high net primary production, which caused a steady increase in the soil organic matter. After 7 years of simulation the old forest detrital matter production decreased due to growth-limiting factors and nutrient limitations. The young forest became increasingly phosphate limited, which resulted in a 10-year decrease in NPP and NLOM. This supports the notion that deciduous forests use more nutrients and are leakier during early stages in forest growth as compared to pines, which are more conservative in their use of nutrients and thus have a competitive advantage (Gholz et al., 1994; Reich et al., 1992).

Table 9.2 Model Testing for Cattail Creek Subwatershed and Comparison to the HSPF Model Statistics

	Data	Model	% Error	HSPF % Error
Total flow	2510.41	2527.58	0.68	8.2
max 10%	930.2	925.79	20.48	4.9
min 50%	587.3	596.25	1.50	214.7
Total 1986	326.16	282.24	215.56	
Total 1987	472.83	469.25	20.76	
Total 1988	482.01	414.22	216.37	20.7–1 18.1
Total 1989	660.62	748.29	11.72	
Total 1990	568.78	611.31	6.96	

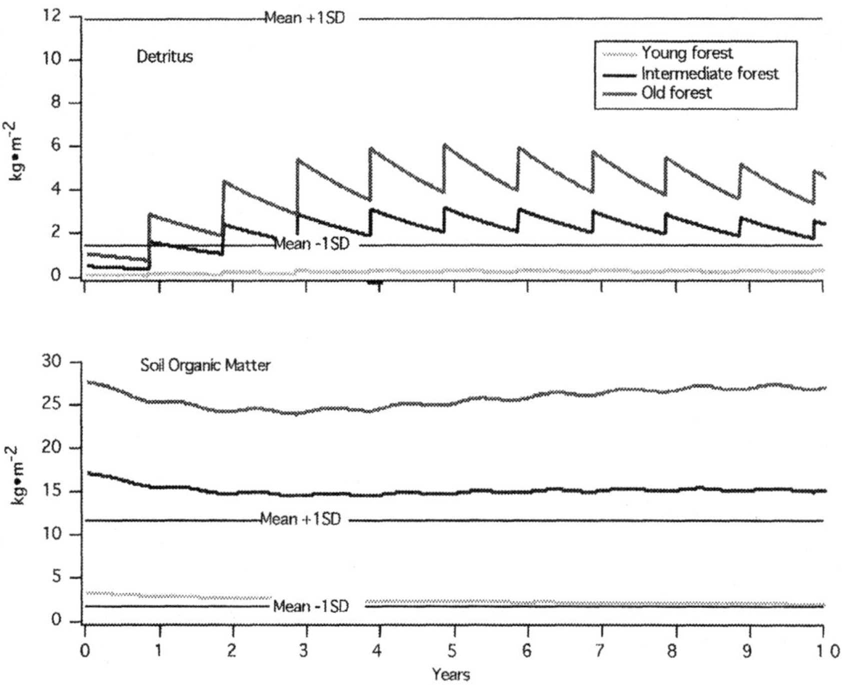

Figure 9.8 Daily output for the detritus and soil organic matter stocks from calibration runs simulating 10 years of forest growth in three successional stages (young, intermediate, and old). Reference points (Mean + SD and Mean − SD) were derived from data tables published by Johnson and Lindberg, 1992 (forest floor organic matter measurements for detritus, and soil organic matter measurements for soil organic matter).

Nutrient concentrations during the 10-year simulation runs tend to stay within the limits reported in the literature (Table 9.2). The standard deviations in the single age runs were lower than would be expected from a spatial run showing mixed ages in the landscape. The literature values represent means of the more variable mixed ages. The forests in all age stages had transitional loads of DIN and low values of inorganic phosphate. Net primary production through the simulation was approximately twice the NPP derived from the FIA database, but well within the range of 1 SD (Table 9.2). Only one third of the variation in NPP and one tenth in PO^{4-} was captured during the unit model calibration runs. Larger ranges in variation for these two variables is expected when calibrating the spatial model on a larger variety of soil conditions. In contrast, the unit model calibration did account for most of the variation in DIN as variation in the nitrogen cycle is less soil specific and more controlled by atmospheric concentrations of gaseous nitrogen species (Gardner et al., 1996).

The same contrast between PO^{4-} and DIN sources (soil vs. atmospheric) is seen in the seasonal model dynamics (Figure 9.9) of the young forest which show little variation over the year in PO^{4-} concentration relative to DIN. The low biomass of the young forest forces it to rely on the slow release of PO^{4-} from a mineral soil, but DIN remains more readily available through wet deposition. As biomass in the forest increases, nutrient dynamics are more controlled by the mineralization of organic matter and plant uptake rates, which provide a larger pool of available nutrients. This leads to less dramatic

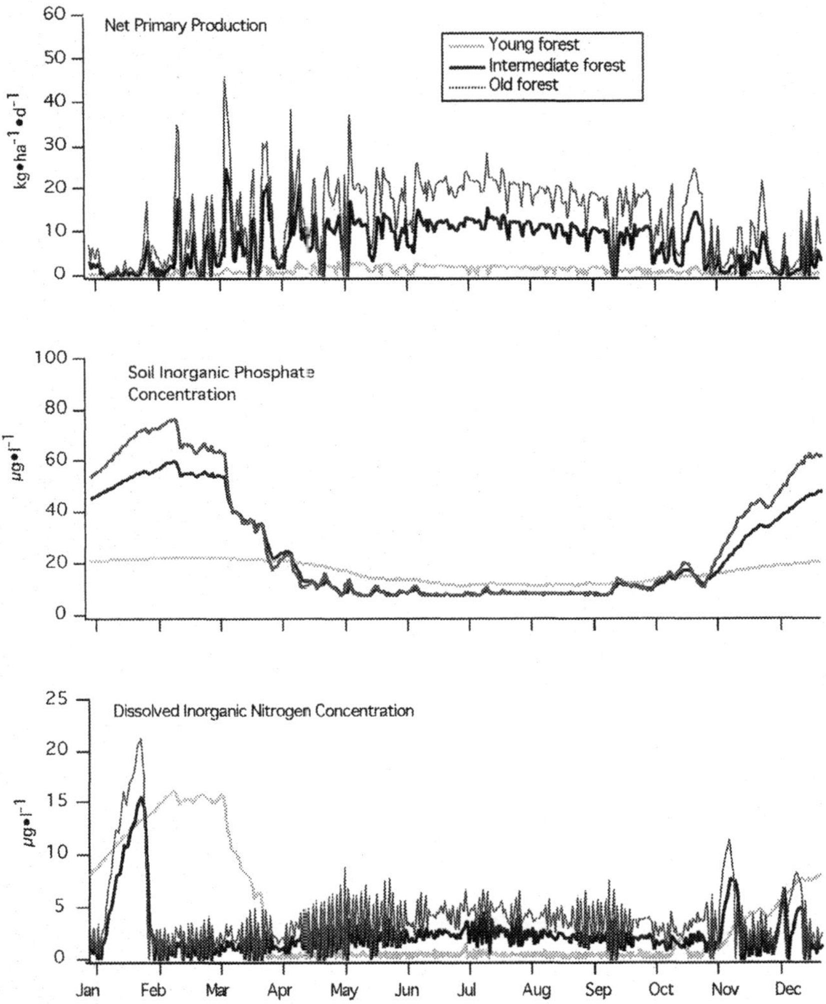

Figure 9.9 Seasonal dynamics for forests of three successional stages (young, intermediate, and old) for net primary production, inorganic phosphate, and nitrogen concentrations. Year 5 of a 10-year calibration run is shown.

changes in nitrogen but more seasonal variation in phosphate in the older forests.

Agricultural land uses were simulated in GEM for crops typical of the Patuxent watershed. Crops included typical cycles between corn, winter wheat, and soybeans. The Pat-GEM was expanded to include decisions on planting, harvesting, and fertilization, while switching between crop-specific growth parameters. Crop production estimates were calibrated against the results from EPIC (Erosion Productivity Impact Calculator) a widely used and calibrated agricultural model. High correlations were found between output generated with Pat-GEM and EPIC. The R^2 varied between 0.87 and 0.98, with results for winter wheat showing the lowest correlation coefficients. Intercalibration using EPIC proved to be a useful "second best" method for calibrating the Pat-GEM model for agricultural land uses. EPIC was able to provide about 30% of the input data required for running the PLM model; and to provide time-series output data (on a daily step) to compare with the GEM output variable. Since EPIC has been widely tested we feel that replicating its results is a sufficient "second best" calibration approach in lieu of detailed local time-series data.

9.3.4 Spatial hydrology calibration

Calibrating and running a hydrologic model of this level of complexity and resolution requires a multistage approach (Voinov et al., 1999a). We first identified two spatial scales at which to run the model—a 200 m and 1 km cell resolution. The 200 m resolution was more appropriate to capture some of the ecological processes associated with land use change but was too detailed and required too much computer processor time to perform the numerous model runs required for calibration and scenario evaluation for the full watershed. The 1 km resolution reduced the total number of model cells in the watershed from 58,905 to 2352.

Second, we identified a hierarchy of subwatersheds. The Patuxent watershed has been divided into a set of nested subwatersheds to perform analysis at three scales (Figure 9.10). The small (23 km^2) subwatershed of Cattail Creek in the piedmont northern part of the Patuxent basin was used as a starting point. Another small subwatershed, Hunting Creek, was selected to calibrate the model for the different hydroecological conditions of the coastal plain area. The next larger watershed was the upper nontidal half of the Patuxent watershed that drained to the USGS gauge at Bowie (940 km^2). And finally we examined the whole Patuxent watershed (2352 km^2). The number of total model cells grew from 566 cells initially, to 23,484 cells for the half watershed, and then to 58,905 cells for the entire study area at the 200 m resolution.

In this stage of the calibrations, we ran only the hydrologic component of the model, without links to the plants and nutrients. While transpiration by plants and the influence of nutrients on plant productivity and transpiration are obviously important influences on hydrology, we excluded them at this stage: (1) for simplicity; (2) for direct comparison to other hydrologic models;

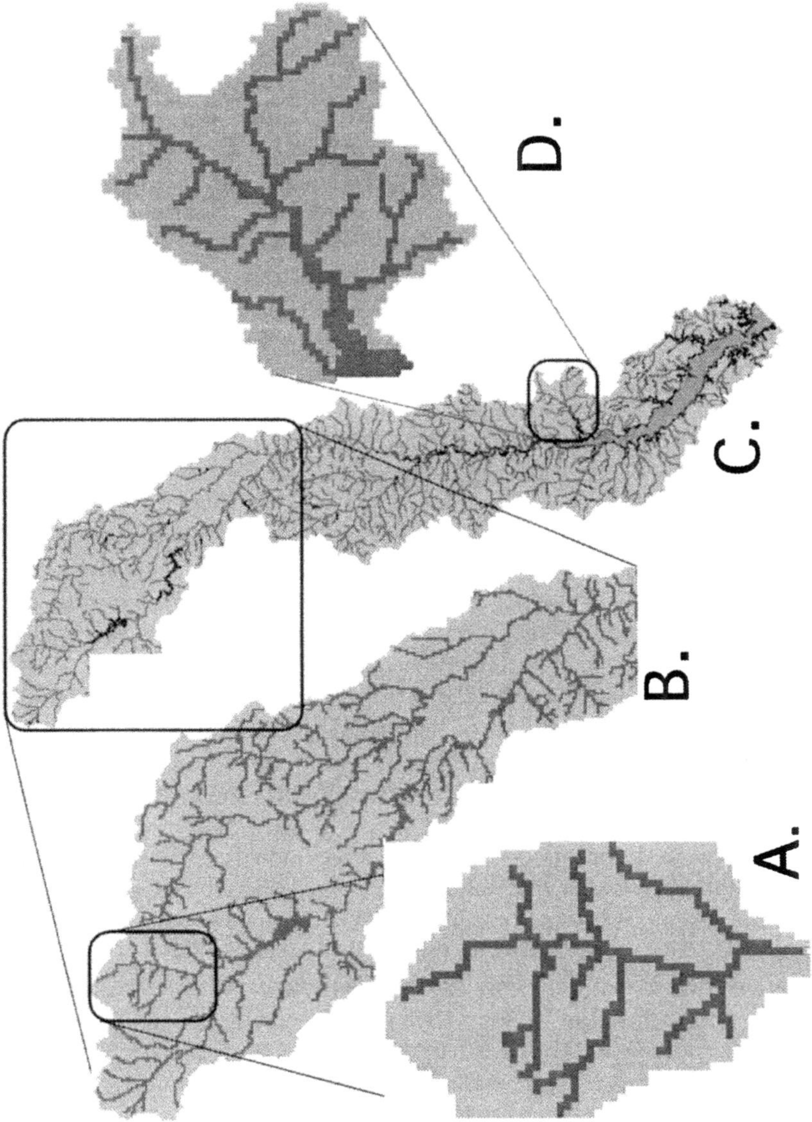

Figure 9.10 Hierarchy of subwatersheds on the Patuxent drainage basin used to calibrate and analyze the model. A. 22.5 km^2 Cattail Creek; B. 939.9 km^2 Upper Patuxent draining at Bowie; C. Full study area; D. Hunting Creek.

and (3) to test the effects of adding the plant and nutrient dynamics later (see below).

We staged a set of experiments with the small Cattail creek subwatershed to test the sensitivity of the surface water flux. Three crucial parameters controlled surface flow in the model: infiltration rate, horizontal conductivity and number of iterations per time step of the unit model. Riverflow peak height was strongly controlled by the infiltration rate. The conductivity determined river levels between storms and the number of iterations modified the width of the storm peaks.

Surface water flow was calibrated against the 13 USGS gauging stations in the area that have data concurrent with the climatic data series (1990–96). In this calibration, model results for the Cattail subwatershed (Figure 9.10a) are in fairly good agreement with the data (Figure 9.11) and may be considered a partial model verification because none of the parameters had been changed after the initial calibration using 1990 data. Some of the flow statistics are presented in Table 9.2, where we have also included calibration results from the Hydrologic Simulation Program—Fortran (HSPF) (Donigian et al., 1984), that has been previously applied to the Patuxent watershed (AQUA TERRA, 1994). We attained a considerably better fit to the data with our model, than with HSPF. HSPF is a more empirically based model that uses high temporal resolution input data (e.g., hourly rainfall data), but relatively low spatial resolution (e.g., aggregated subwatersheds). Much more of the behavior in HSPF is embedded in the parameters of the model than in the PLM (which uses the pattern of land use to drive much of the behavior). Thus the effects of changes in the spatial pattern of land use (one of our key questions) cannot be adequately addressed using HSPF, since it would require recalibrating the model for the new land use pattern, and empirical hydrologic data for this new, hypothetical, land use pattern obviously does not exist.

Next we performed a spatial scaling experiment which involved varying the spatial resolution of the model. Using GIS operations we aggregated the input maps and the model, switching from a 200 m to a 1 km cell resolution. Model runs for the 1 km resolution were remarkably close to the results from the 200 m model (Figure 9.12). This finding was especially promising for analysis of the other modules of the full ecological economic model. Since most of the horizontal spatial dynamics is governed by the hydrologic fluxes, the coarser 1 km resolution should be sufficient for the spatial analysis of the integrated model of the watershed in this case.

Overall, the model of the half watershed (Figure 9.10B) showed less precise model results (Figure 9.12) than the Cattail subwatershed, as predicted from theory (Costanza and Maxwell, 1994). The calibration statistics for the half watershed area are summarized in Table 9.3. In general, the increased spatial extent of the model presented more heterogeneity, which was not fully accounted for in the calibration. Specific reasons for this include the spatial resolution of the input climatic data, the greater complexity of ground-

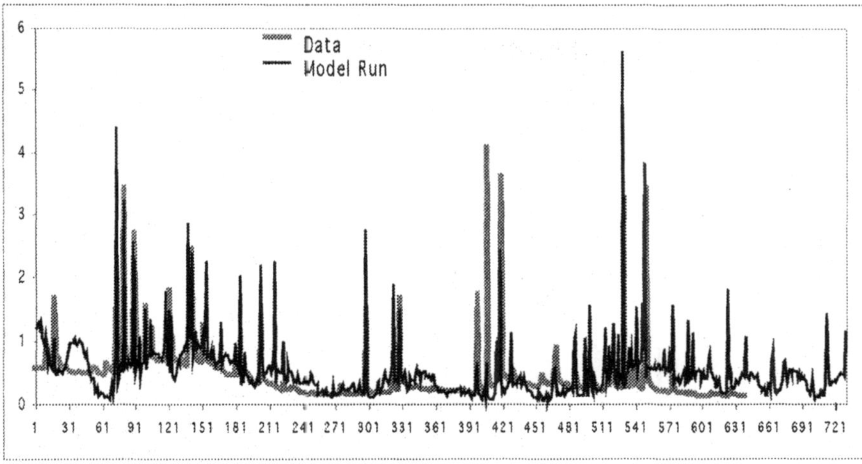

A.

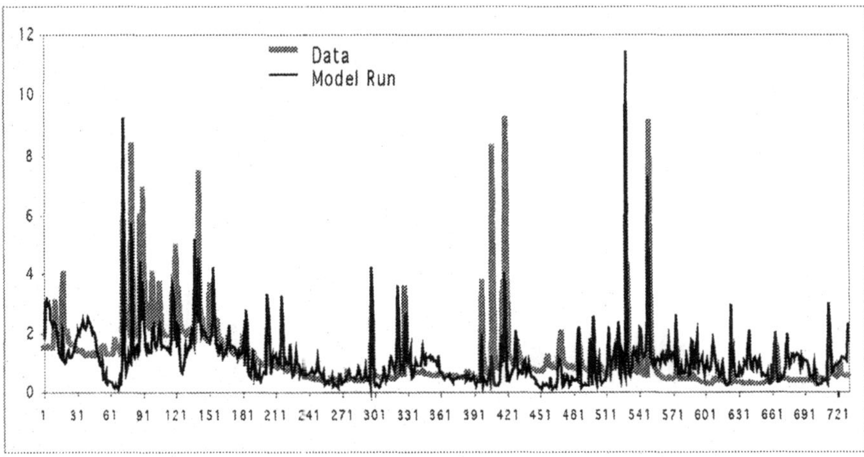

B.

Figure 9.11 Calibration (first 365 days) and verification (second 365 days) of the spatial hydrologic module based on the 1980–81 data for two gaging stations on Cattail Creek. A. Cooksville station; B. Glenwood station.

water flows at this scale (which are handled in a very simplified way in the model), and the need to recalibrate some of the hydrologic parameters at the larger scale.

One parameter that needed to be adjusted was the number of iterations *N* in the hydrologic module. At the larger watershed scale, it turned out that a better fit could be obtained if the number of iterations was further increased. Apparently this was because at this larger scale we needed to

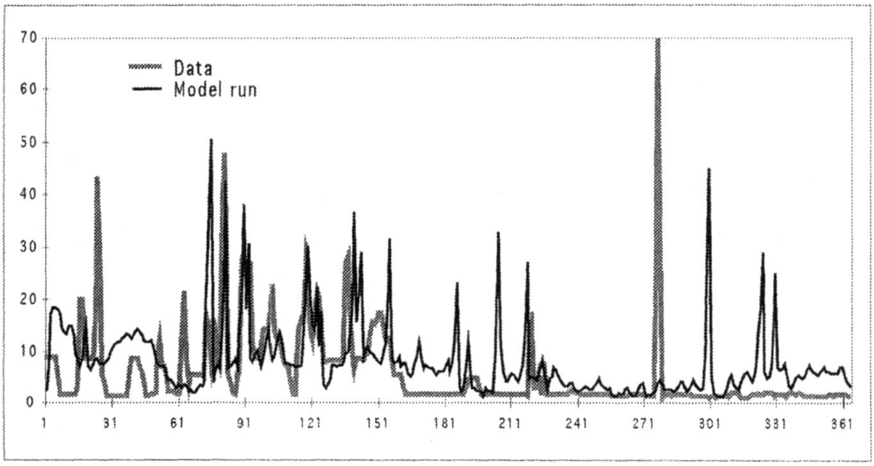

A.

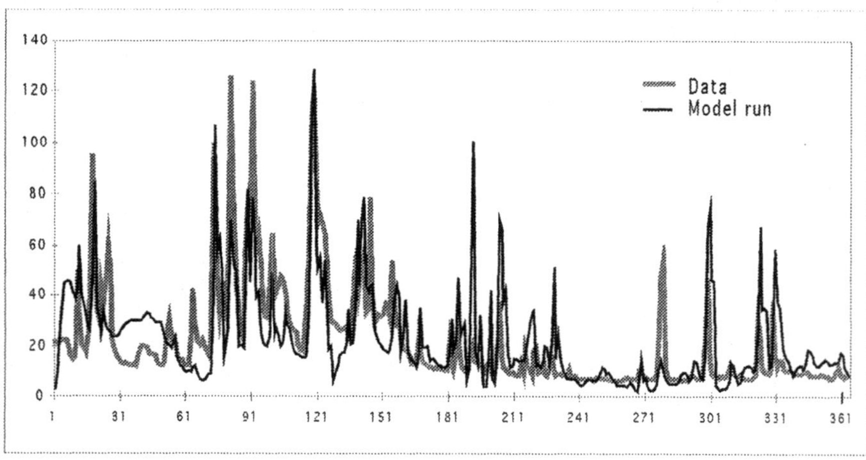

B.

Figure 9.12 Verification for the spatial hydrologic module based on the 1980–81 data for two gauging stations in the upper subwatershed. A. Laurel station. This station is located immediately after a reservoir, which the operation schedule is not accounted for in the model. This explains the flat baseflow rate measured in summer, as well as the high flow on day 275 caused by opening the tainter gate in the dam. B. Bowie station.

move water farther and faster to better simulate the short-term high peaks that were observed in the data. This was a clear illustration of the fact that different scales present new emerging behavior of the system, and that rescaling

Table 9.3 Model Testing and Comparison to the HSPF Model Statistics for the Half Subwatershed Draining at Bowie and the Unity Subwatershed

	Unity				Bowie			
	Data	**Model**	**% error**	**HSPF % error**	**Data**	**Model**	**% error**	**HSPF % error**
Total flow	3950.54	3981.31	0.8	−2.1	36617.43	37978.78	3.6	9.7
max 10%	1410.15	2148.13	41.5	2.3	12497.58	16546.70	27.9	15.1
min 50%	826.76	626.78	−27.5	−12.1	7917.98	6582.62	−18.4	9.0
Total 1986	484.52	446.30	−8.2		4752.94	4352.84		
Total 1987	816.48	942.00	14.3		6446.08	7041.22	8.8	
Total 1988	819.30	792.10	−3.4	−11.6 to +8.0	6751.99	5841.62	−14.5	−2.6 to +25.3
Total 1989	960.30	949.45	−1.1		10507.98	11881.88	12.3	
Total 1990	869.94	851.47	−2.1		8158.45	8861.23	8.3	

is an important process that can not usually be done mechanically. The best fit to data was obtained when running the model with the self-adjusting method for N with the maximum number of iterations $m = 100$ (Voinov et al., 1999a). Interestingly, the Cattail subwatershed still performed as well as before with this value of m. This could be expected since the previous analysis showed that there was no sensitivity of subwatersheds to increases in N beyond 20 ($m = 20$).

Within the subwatershed we assumed that the groundwater dynamics closely follow the surface water flows and confined the groundwater to the subwatershed area. This is probably not accurate even for Cattail Creek and at larger scales the groundwater patterns are even more complex.

The spatial rainfall and other data were interpolated from daily records of seven stations distributed over the study area. The Cattail Creek subwatershed hydrology was driven by one climatic station and the half-watershed model incorporated data from three stations. The lack of data on the true variability of the meteorological data in space and time hinders the model's ability to accurately represent short term or localized responses in river flow. However, the model is able to consider antecedent moisture and runoff-generating areas in a spatially realistic manner based on topography, land use, and soil type, giving the simulation a fairly high degree of precision. The general hydrologic trends seem to be well captured by the model and therefore allow us to expand the study to other modules. We also refer the reader to our web page at http://iee.umces.edu/PLM, which further describes the model.

9.3.4.1 Full ecological model calibration

The full spatially explicit ecological model was run for several years using historical climate inputs for calibration purposes. In this case we ran the model at a 1 km spatial resolution. We used two methods to compare the model performance to the available data.

Certain modeled variables or indices that aggregate model variables were compared to point time series data such as streamflow, nutrient concentration in the streams, and historical tree-ring data for the region (Table 9.1). The inclusion of plant and nutrient dynamics improved the model's hydrologic performance in comparison to the output reported above. The spatially explicit representation of plant and nutrient dynamics modifies the evapotranspiration and interception fluxes in the model, making the model performance more realistic. It was also essential for scenario runs that take into account land use and cover changes, in which these changes modify the hydrologic fluxes in the watershed.

A sample of calibration for flow weighted nitrogen concentrations in the Patuxent River at Bowie is presented in Figure 9.13. Data is available to calibrate longer-term runs of the model with these data sets. Model output was compared to field data by visually inspecting superimposed graphs and comparing annual mean and total values. For example, the long–term trend of nitrogen dynamics in Hunting Creek—a small subwatershed in the Coastal

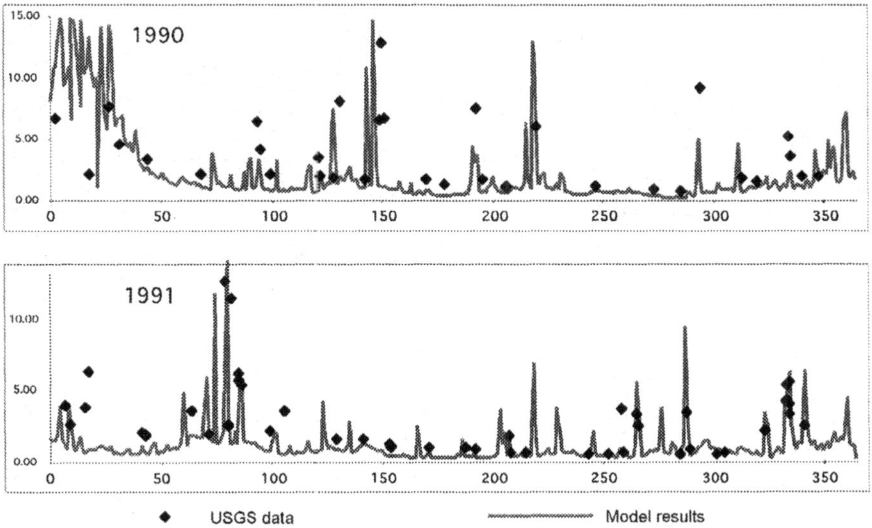

Figure 9.13 Results of calibration for flow weighted nitrogen concentrations in the Patuxent River at Bowie. A. 1990; B. 1991.

Plain area—shows good correlation with the annual dynamics calculated from the data (Figure 9.14).

Comparison of raw spatial data is a much more difficult and less studied procedure. Spatially explicit data are scarce and rarely match the spatial extent and resolution produced by the model. Some example output for plant primary production from the model is shown in Figure 9.15. This shows the typical pattern of seasonal growth in the region, which has a significant impact on hydrology through transpiration. Data derived from AVHRR satellite images, the NDVI or "greenness" index, were used to calibrate the full model's predictions of primary production for intra-annual effects. We created an index from the NDVI data so that we were comparing the magnitude of NDVI change to the magnitude of NPP change between cells in time and space. A visual comparison shows fairly good agreement between the model output and the data currently available. For long-term forest growth calibrations we used the FIA data on forest growth (Table 9.4). The small Cattail Creek subwatershed (which can be run more quickly) was run for 50 years to examine long-term trends. These results compared favorably to the FIA data.

9.4 Scenarios

The goal of the linked ecological economic model development was to test alternative scenarios of land use patterns and management. A wide range of future and historical scenarios may be explored using the calibrated model. We have developed scenarios based on the concerns of county, state, and

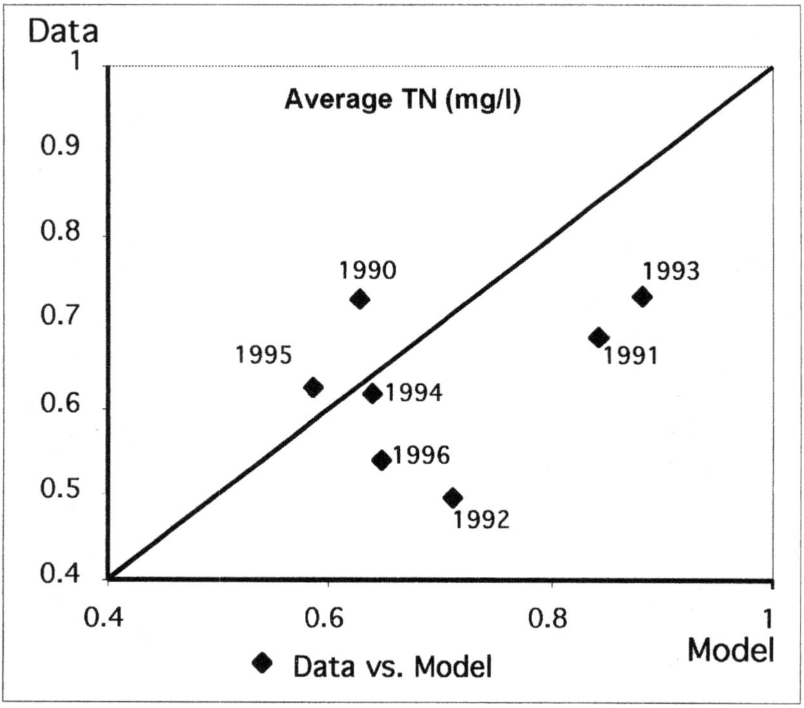

Figure 9.14 Long-term annual dynamics of total nitrogen in Hunting Creek.

federal government agencies, local stakeholders, and researchers. The following set of 18 initial scenarios were considered:

- A group of **historical scenarios** based on the USGS reconstruction (Buchanan et al., 1998) of land use in the Patuxent watershed:

Table 9.4 10-year Averages for Three Forest Model Variables Compared to Literature Values

	NPP (kg/m^2/y)	PO^{4-}(µg/l)	DIN (µg/l)			
	Mean	**SD**	**Mean**	**SD**	**Mean**	**SD**
		Model output				
Young	0.039	0.006	0.017	0.004	4.1	5.5
Intermediate	0.29	0.014	0.025	0.019	2.7	2.6
Old	0.497	0.014	0.031	0.027	4.2	3.5
All forest ages	0.27	0.190	0.024	0.02	3.7	4.1
Reference data						
All forest ages	0.14[a]	0.67	0.185[b]	0.165	5[c]	5

[a]Derived through the FIA Database for the Patuxent watershed.

[b]Midoint and maximum deviation reported by Stevenson (1986) for sandy soils.

[c]Midpoint and maximum deviation reported by Aber (1996) for deciduous forests.

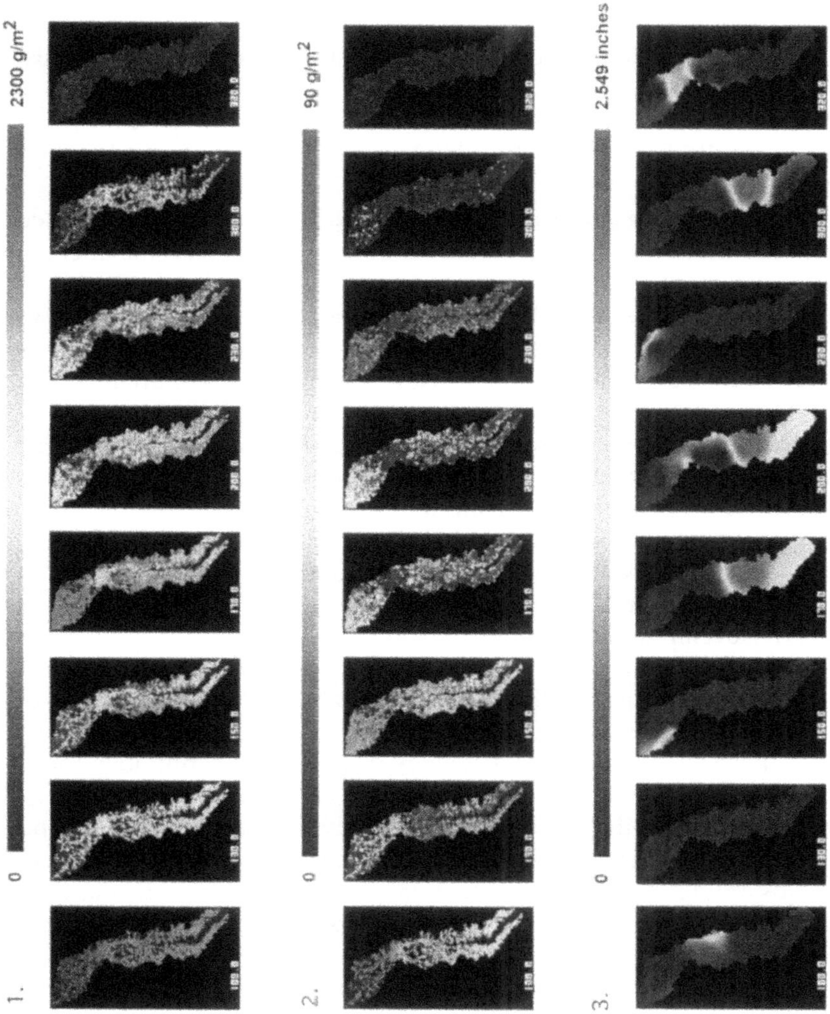

Figure 9.15 Sample spatial output of the full model 1. Photosynthetic biomass 2. Net primary production 3. Rainfall.

—(1) 1700: predevelopment era. Most of the area forested, zero emissions.
—(2) 1850: agrodevelopment. Almost all the area under agricultural use, traditional fertilizers (marl, river mud, manure, etc.), low emissions.
—(3) 1950: decline of agriculture, start of reforestation and fast urbanization.
—(4) 1972: maximal reforestation, intensive agriculture, high emissions.

- (5) **Baseline scenario.** We use 1990 as a baseline to compare the modeling results. The 1990–1991 climatic patterns and nutrient loadings were used.
- (6) **1997 land use pattern.** This data set has just recently been released and we used it with the 1990–1991 forcings to estimate the effect of land use change alone.
- (7) **Buildout scenario.** With the existing zoning regulations, we assumed that all the possible development in the area occurred. This may be considered as the worst-case scenario in terms of urbanization and its associated loadings.
- (8) **Best management practices (BMP)**—1997 land use with lowered fertilizer application and crop rotation. These management practices were also assumed in the remaining scenarios.
- A group of scenarios of change in land use over the 5 years following 1997 (i.e., for 2003) based on the **economic land use conversion (ELUC)** Model by N. Bockstael:
  —(9) Development as usual.
  —(10) Development with all projected sewer systems in place.
  —(11) Development with no new sewers but contiguous patches of forest 500 acres and more protected.
  —(12) Development with all sewers in place and contiguous forest protected.
- A group of hypothetical scenarios to study **dramatic change** in land use patterns using the 1997 land use as the starting point. These scenarios are designed to show the total contribution of particular land use types to the current behavior of the system by completely removing them.
  —(13) Conversion of all current agricultural land into residential land.
  —(14) Conversion of all current agricultural land into forested land.
  —(15) Conversion of all current residential land into forested land.
  —(16) Conversion of all current forested land into residential land.
- Another group of hypothetical scenarios to study the **effects of clustering,** again using the 1997 land use as the starting point:
  —(17) Residential clustering—conversion of all current low-density residential land use into urban land use around three major centers.

—(18) Residential sprawl—conversion of all current high-density urban land use into residential land use randomly spread across the watershed.

The scenarios were driven by changes in the land use map, the sewers map, patterns of fertilizer application, amounts of atmospheric deposition, and location and number of dwelling units. Since the model is spatially explicit and dynamic, it generates a huge amount of output for each scenario run. We can only present a brief summary here in the form of spatially and temporally averaged values for a few key indicators. Table 9.5 is a summary of some of the model output from the different scenarios looking at nitrogen concentration in the Patuxent River as an indicator of water quality, changes in the hydrologic flow, and changes in the net primary productivity of the landscape. Some selected additional results of the scenario runs are described briefly below.

9.4.1 Historical scenarios

In this group of scenarios we attempted to reconstruct the historical development of the Patuxent watershed, starting from the pre-European settlement conditions in 1700. The 1850, 1950, and 1972 maps (Figure 9.16) were produced based on data from Buchanan et al. (1998). In 1700 the watershed was almost entirely forested, with very low atmospheric deposition of nitrogen, no fertilizers, and no septic tank discharges. The rivers had very low nutrient concentrations. By 1850 the landscape had been dramatically modified by European settlers. Almost all the forests were cleared and replaced with agriculture (Table 9.5). However, fertilizers used at the time were mostly organic (manure, river mud, green manure, vegetable matter, ashes), the atmospheric deposition of nutrients was still negligible, and the population was low, producing little septic tank load.

After 1850, agricultural land use began shrinking and forests began regrowing. By 1950 the area of forests had almost doubled. At the same time, rapid urbanization began, primarily along the Washington, D.C.–Baltimore corridor. This affected the Patuxent watershed both directly (through changes in land use from agriculture and forests to residential and commercial uses) and indirectly (through increased auto use in the larger region and increased atmospheric inputs of nutrients). This process continued until the 1970s, when reforestation hit its maximum. Since then, continued urbanization of the area has been affecting both agricultural and forested areas at approximately the same rate. The atmospheric emissions and fertilizer applications were assumed to grow steadily from the low preindustrial levels to modern load levels. The growing population in the residential sectors was contributing to growing discharges from septic tanks.

9.4.2 1990 vs. 1997 vs. buildout

Comparison between the 1990 and 1997 conditions shows that there was a considerable decline in the number of forested and agricultural cells, which

Table 9.5 **Some results of scenario runs for the Patuxent model.** The buildout conditions (LUBO) were estimated based on the existing zoning maps and the average population densities for particular land use types. The buildout conditions represent the worst case scenario. The ELUC scenarios (LUB1-4) are based on the model by N. Bockstael. The historical scenarios (LU1700-1972) are a reconstruction based on estimates done by USGS. Total NPP ($g/m^2/day$) presents the average across the whole watershed productivity of the plan ecosystem. It well represents the approximate proportion of forested and agricultural land use types, which have a larger NPP than the residential ones. N gw.c.—is the average concentration of Nitrogen in the groundwater. Since groundwater is a fairly slow variable in the model and the model had only 1 year of relaxation time in the experiments performed, it is most likely that this parameter does not adapt fast enough to track the changes assumed under different scenarios. Wmax is the total of the 10% of the flow that is maximal over a one year period. This represents the peak flow. Wmin is the total of the 50% of flow that is minimal over a 1-year period. This is an indicator of the baseflow.

		Forest	Resid	Urban	Agro	Atmos	Fertil	Decomp
			number of cells			kg/ha/year		
1	LU1700	23860	0	0	56	2.67	0.00	253.03
2	LU1850	348	7	0	2087	5.35	93.15	123.83
3	LU1950	911	111	28	1391	96.27	113.31	144.66
4	LU1972	1252	223	83	884	85.58	156.21	175.07
5	LU1990	1315	311	92	724	80.23	114.58	164.57
6	LU1997	1195	460	115	672	80.23	112.75	150.34
7	LUBO	312	729	216	1185	85.58	184.34	74.72
8	BMP	1195	460	115	672	80.23	61.14	170.03
9	LUB1	1129	575	134	604	85.58	97.14	145.45
10	LUB2	1147	538	134	623	85.58	98.03	148.19
11	LUB3	1129	577	134	602	85.58	96.96	145.37
12	LUB4	1133	564	135	610	85.58	97.55	146.12
13	S5a2r	1195	1132	115	0	85.58	47.26	127.65
14	S6a2f	1867	460	115	0	85.58	19.19	197.63
15	S6r2f	1655	0	115	672	85.58	82.14	203.00
16	S7f2r	0	1655	115	672	85.58	143.07	30.40
17	S8clust	1528	0	276	638	85.58	78.36	187.85
18	S9spra	1127	652	0	663	85.58	106.77	147.81

Table 9.5 **Some results of scenario runs for the Patuxent model.** (cont.)

		Septic	N aver	N max mg/l	N min	Wmax	Wmin m/year	N gw c mg/l	NPP kg/m2/d
1	LU1700	0.09	4.14	47.38	0.06	71.934	31.493	1.356	50.308
2	LU1850	3.54	5.65	51.0	0.25	71.849	27.797	1.675	15.739
3	LU1950	9.44	12.03	124.53	0.32	74.112	26.447	1.710	26.586
4	LU1972	9.60	18.29	281.36	0.28	78.711	23.556	1.723	33.568
5	LU1990	15.49	10.82	168.95	0.10	83.081	21.191	1.676	31.670
6	LU1997	22.81	10.28	106.86	0.14	85.312	19.624	1.691	28.974
7	LUBO	26.22	12.02	159.00	0.28	97.957	15.027	1.862	12.713
8	BMP	15.49	6.50	58.47	0.16	83.307	21.694	1.621	30.680
9	LUB1	29.45	8.48	68.30	0.32	85.414	19.631	1.709	26.323
10	LUB2	16.03	8.55	66.52	0.24	85.269	19.667	1.687	26.820
11	LUB3	29.54	8.43	66.72	0.32	85.417	19.654	1.709	26.315
12	LUB4	17.17	8.56	66.62	0.20	85.250	19.494	1.689	26.438
13	S5a2r	52.70	6.71	52.84	0.29	83.069	13.182	1.651	25.597
14	S6a2f	21.78	7.25	45.66	0.24	84.928	18.857	1.561	39.67 7
15	S6r2f	5.50	7.82	91.15	0.25	87.805	28.359	1.617	37.289
16	S7f2r	75.989	9.40	170.17	0.29	89.734	10.004	1.922	3.477
17	S8clust	13.02	8.87	91.66	0.39	121.236	24.621	1.633	34.241
18	S9spra	26.61	7.05	35.48	0.30	70.130	20.679	1.718	26.441

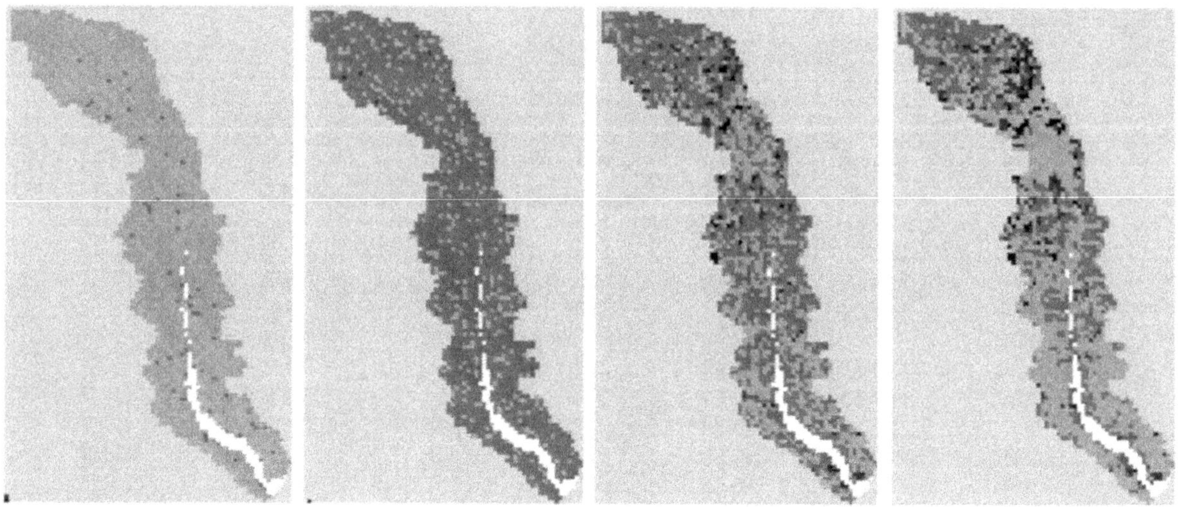

	1700	1850	1953	1972
Urban	0	0	28	56
Residential	0	7	111	195
Agro	56	2087	1391	800
Forest	2386	348	911	1391
Water	118	118	118	118

Figure 9.16 Approximate reconstruction of Patuxent watershed development in 1700–1972, based on USGS estimates (Buchanan et al., 1998).

was due to the increase in residential and urban areas. Accordingly, fertilizers contributed less to the total nitrogen load for the watershed, whereas the amount of nitrogen from septic tanks increased (Table 9.5). These load totals also demonstrate the relative importance of different sources of nitrogen on the watershed. Under existing agricultural practices the role of fertilizers remains fairly high. Atmospheric deposition contributes unexpectedly high proportions of the nitrogen load. The role of septic tanks may seem minor, but it should be remembered that the fate of septic nitrogen is quite different from that of fertilizer and atmospheric nitrogen. Under the existing design of septic drainage fields, the septic discharge is channeled directly to groundwater storage almost entirely avoiding the root zone and nutrient uptake by terrestrial plants.

From 1990 to 1997 most of the land use change occurred by replacing forested with residential land use types. As a result we do not observe any substantial decrease in water quality in the Patuxent River (Table 9.5). The changes in hydrologic parameters that are associated with the substitution of residential areas for forested and agricultural ones result in somewhat more variability in the flow pattern; however, this difference is not very large. Apparently during this time period, residential land use is still less damaging than agricultural use and the loss in environmental quality that is associated with a transfer from forested to residential conditions is compensated by a net gain in a similar transfer from agricultural to residential use.

These trends are reversed when we move on to the buildout (BO) conditions. At some point, a threshold is passed after which most of the development occurs due to deforestation and the effect of residential and urban use becomes quite detrimental for the water quality and quantity in the watershed. The base flow (represented by the 50% minimal flow values) decreases to almost half of the predevelopment 1700 conditions, and the peak flows become very high because of the overall increase of impervious surfaces. Accordingly, the nitrogen content in the river water grows quite considerably.

9.4.3 Best management practices

The next scenario attempts to mimic the possible effects of BMPs. Government concerns are primarily aimed at nutrient reduction through non-point-source control and growth management (MOP/MDE, 1993) with the broader goal of improving the groundwater, river, and estuarine water quality for drinking water and habitat uses. Non-point-source control methods under study include stream buffers, adoption of agricultural and urban BMPs, and forest and wetland conservation. Urban BMPs or storm water management involve both new development and retrofitting older developments. Growth management includes programs to cluster development, protect sensitive areas, and carefully plan sewer extensions. Clustered development has been proposed and promoted in Maryland as a method to reduce non-point sources and preserve undeveloped land.

At this time we have limited our consideration of BMPs to reduction of fertilizer application. Crop rotation has been assumed previously as a standard farming practice in the area. In addition, we assessed the potential for nutrient reduction in the Patuxent from reductions achieved by farmers in the basin who have adopted farm nutrient management plans. The Maryland Nutrient Management Program (NMP) enlists farmers who are willing to create and implement nutrient management plans that use a variety of techniques to lower application rates including nutrient crediting with and without soil testing, setting realistic yield goals, and manure testing and storage. The biggest gains for farms without animal operations tended to come from adjusting yield goals (Steinhilber, 1996). From this information, we created an expected nutrient reduction of 10 to 15%, which is the typical reduction for farms in the NMP (Simpson, 1996). Another major source of fertilizer application reduction is accounting for atmospheric deposition in calculations of nutrient requirements. This has been promoted by some of the recent recommendations issued by MDA. As a result, we get quite a considerable change in fertilizer loading and reduction of agricultural land use in the watershed no longer becomes beneficial for water quality in the river (Table 9.5).

9.4.4 Economic land use conversion model scenarios

This group of scenarios distributes 28,000 projected new dwelling units (using 1997 conditions as a base) within the area of the seven counties that include the Patuxent watershed under certain assumptions about the location of sewers and forest preservation strategies. Most of the change occurs in the upper Patuxent portion of the watershed. As seen from Table 9.5 the resulting changes in land use distributions were not as dramatic as during the 1990–97 period. Correspondingly, the changes in water quality in the river were quite subtle. Our indicators show less than a 1% change relative to the 1997 conditions. However, it is noteworthy that in these scenarios contrary to the previous period most of the land use change is from agricultural to residential habitats. The reduction of agricultural loadings turns out to be more important than the increase in septic tank discharges. Because of the high primary productivity of agricultural land use relative to residential, we also observe a decline in average NPP. Apparently these changes do not bring us to the threshold conditions after which the residential trends of development become especially damaging to the environmental conditions.

9.4.5 Hypothetical scenarios

In the next group of scenarios we considered some more drastic changes in land use patterns. None of these is realistic, but they allow one to estimate the relative contributions of major land use types to the current behavior of the system. For example, by comparing Scenarios 14 and 15 one can see that agricultural land uses currently play a larger role in the nutrient load received by the river than residential land uses, even under the BMPs. We get a considerable gain in water quality by transferring all the agricultural land into residential. Contrary to expectations, cluster development (Scenario 17) did not turn out to be any better for river water quality than residential sprawl (18). Because of larger impervious areas associated with urban land use, the peak runoff dramatically increased in this scenario. This in turn increased the amount of nutrients washed off the catchment area. Cluster development would be beneficial only if it is accompanied by effective storm water management that will reduce runoff and provide sufficient retention volumes to channel water off the surface into the groundwater storage.

Conversion of all currently forested areas into residential (Scenario 16) was almost as bad as the buildout scenario (7). However, the crop rotation assumed in Scenario 16 decreased the amounts of fertilizers applied somewhat and resulted in lower overall nitrogen concentrations. The septic load in this case was so large because the transition to residential land use was assumed to occur without the construction of sewage treatment plants. In the buildout scenario most of the residential and urban dwellings were created in areas served by existing or projected sewers.

9.4.6 *Summary of scenario results*

One major result of the analysis performed thus far is that the model behaves well and produces plausible output under significant variations in forcing functions and land use patterns. It can therefore be instrumental for analysis and comparisons of very diverse environmental conditions that can be formulated as scenarios of change and further studied and refined as additional data and information are obtained. The real power of the model comes from its ability to link spatial hydrology, nutrients, plant dynamics, and economic behavior via land use change. The economic submodel incorporates zoning, land use regulations, and sewer and septic tank distribution to provide an integrated method for examining human response to regulatory change. The projections from the economic model of land use change based on proposed scenarios shows the probable distribution of new development across the landscape so that the spatial ecological aspects can be evaluated in the ecological model. The model allows fairly site specific effects to be examined as well as regional impacts so that both local water quality and Chesapeake Bay inputs can be considered.

The scenario analyses also demonstrated that the Patuxent watershed system is complex and its behavior is counterintuitive in many cases. For example, in the entirely forested watershed of 1700, the flow was very well buffered showing very moderated peaks and fairly high base flow. The agricultural development that followed in the next century actually decreased both the peak flow and the base flow, contrary to what one would expect, even though the decrease in the base flow was much more significant than the decrease in the peaks. Apparently, evapotranspiration rates for the kinds of crops currently included in the model was high enough to keep the peaks down. Comparing the effects of various land use change scenarios on the water quality in the river (Figure 9.17) similarly shows that the connection

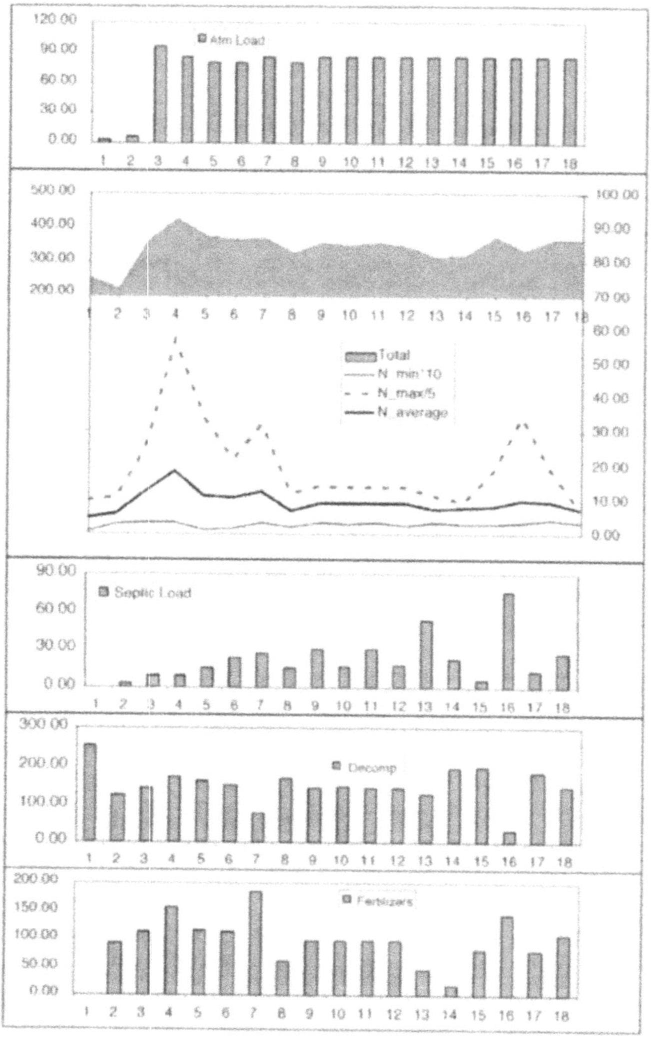

Figure 9.17 Nitrogen loading and concentration of nitrogen in the Patuxent River under different scenarios of land use.

between the nutrient loading to the watershed and the nutrient concentration in the river is complex and difficult to anticipate or generalize. This merely confirms the need for a complex, spatially explicit simulation model of the type we have developed here. Nevertheless, a few general patterns emerge from analysis of the scenario results, including:

- As previously observed (Krysanova et al., 1999), the effects of temporarily distributed loadings are less pronounced than event-based ones. For example, fertilizer applications that occur once or twice a year increase the average nutrient content and especially the maximum nutrient concentration quite significantly, whereas the effect of atmospheric deposition is much more obscure. The difference in atmospheric loading between Scenarios 1 and 3 is almost two orders of magnitude, yet the nutrient response is only five to six times higher, even though loadings from other sources also increase. Similarly, the effect of constant increases in septic loadings is not so large. The average N concentration is well correlated (corr = 0.77) with the total amount of nutrients loaded. The effect of fertilizers is also high (corr = 0.74), while the effect of other sources is much less (septic corr = 20.0075; decomposition corr = 20.2267; atmosphere corr = 0.49). The fertilizer application rate determines the maximum nutrient concentrations (corr = 0.71), with the total nutrient input also playing an important role (corr = 0.60). Even the groundwater concentration of nutrients is closely related to fertilizer applications (corr = 0.89); however, in this case the septic loadings also play an important role (corr = 0.59), even a more important one than the total N loading (corr = 0.44).

- The hydrologic response is driven strongly by the land use patterns. The peak flow (max 10% of flow) is almost entirely determined by the degree of urbanization (corr = 0.94). The base flow (min 50% of flow) is somewhat correlated with the number of forested cells (corr = 0.54), but there are obviously other factors involved.
- Different land use patterns result in quite significant variations in the net primary productivity (NPP) of the watershed, both in the temporal (Figure 9.18) and in the spatial (Figure 9.19a and b) domains. The predevelopment 1700 conditions produce the largest NPP, under buildout conditions NPP is the lowest. In the latter case, the dynamics of NPP were more representative of the agricultural land use with higher NPP values attained later in the year as crops mature. Interestingly, under the BMP scenario with lower fertilizer applications we still get a higher NPP than under reference conditions of 1997. From Figure 9.18 it is clear that this is primarily because of the crop rotation and growth of winter wheat that matures earlier in the season than corn.

Some model results are difficult to interpret. For example, it is not quite clear why the residential sprawl scenario produced a lower peak flow than the 1700 all-forested scenario. In both cases there was no urban land use with its associated large impervious areas. However, one would expect that forests should have a better peak retention capacity than agricultural and residential land uses. These are the types of output that generate questions for further investigation and potential model improvement that can lead to better understanding of the processes involved.

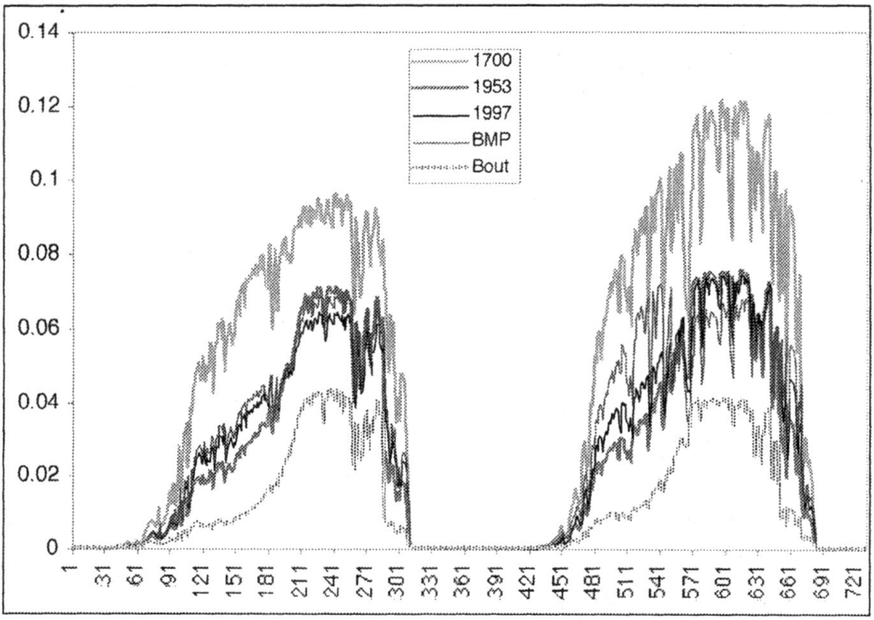

Figure 9.18 Variations in dynamics of NPP under different scenarios.

9.4 Discussion

The linked ecological economic model is a potentially important tool for addressing issues of land use change. The model integrates our current understanding of certain ecological and economic processes to give best available estimates of the effects of spatially explicit land use or land management change. The model also highlights areas where knowledge is lacking and where further research should be targeted.

The high data requirements and computational complexities for this type of model mean that development and implementation are relatively slow and expensive, but for many of the questions being asked this complexity is necessary. We have tried to find a balance between a simple and general model which minimizes complexity and one that provides enough process-oriented, spatially and temporally explicit information to be useful for management purposes. Spatial data is becoming increasingly available for these types of analyses and our modeling framework is able to take advantage of spatial and dynamic data in its relatively raw form without being forced to use complex spatial or temporal aggregation schemes. System dynamics strongly influence ecosystem processes, with processes changing in dominance over time. Our model allows these changes to be incorporated. We will continue to trim model components where possible based on sensitivity and scaling analyses but wish to maintain the generality of the GEM and the spatial explicitness of the PLM.

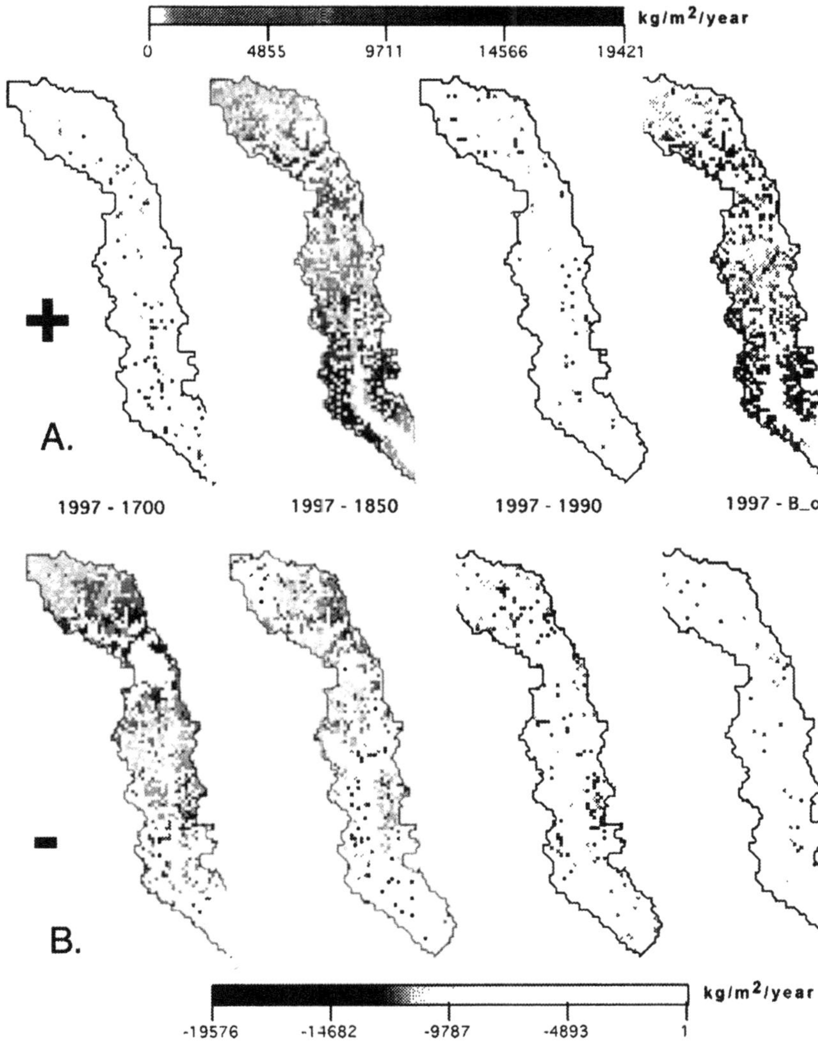

Figure 9.19 Spatial distribution of change in NPP on the Patuxent watershed. a) Increase in NPP; b) Decrease in NPP.

9.4.1 Future work

Future work will consider a range of additional scenarios including climate change scenarios (changing storm frequency and intensity and CO_2 enrichment effects on plants) and additional development patterns that reflect specific "smart growth" initiatives. The model will also be used in "design mode." A series of stakeholder workshops are planned with the goal of achieving broader consensus on the preferred environmental and economic endpoints for the system. The model can be used to both inform this discussion and to determine the best way to achieve the desired endpoints.

Model development will continue with (1) a series of scaling experiments to better understand the trade-offs between spatial, temporal, and complexity resolution and model performance; (2) addition of spatially explicit economic and social component to the unit model that will track built capital, human capital, and social capital; and (3) addition of spatially explicit animal population models for deer, beaver, and other "landscape structuring" species.

We also plan to continue software development to make spatially explicit landscape modeling more accessible. The Modular Modeling Language we have been developing (Maxwell and Costanza, 1995) offers the promise that submodels or modules of varying degrees of detail can be developed independently and interchanged during model development. Then, as users implement a model for a particular area, modules can be selected based on the relative importance of local processes and high detail can be used where needed but otherwise avoided. More work also remains to be done to refine the model and address both data and model scaling issues.

We expect that these efforts will further advance our understanding of the often subtle and indirect links among and between the ecological and economic parts of regional landscape systems so that we may make the complex "place based" decisions we face more wisely.

section four

Conclusions

chapter ten

Future directions

Robert Costanza, Bobbi S. Low, Elinor Ostrom, and James Wilson

Contents

In this book we have presented models at various levels of generality in order to illustrate an accessible approach to the analysis of socio-ecological systems. We proposed a framework that links human decision systems and ecosystems, then explored an increasingly richer and complex set of models that illustrate our human-ecosystem framework. Finally, we reviewed examples of ways such models can be used for analysis, teaching, and consensus

1-5667-0389-1/01/$0.00+$.50

building in the real world. Although we have linked some of our dynamic analytical models to real-world problems, clearly we have only started what needs to be a significant, long-term endeavor.

Much more remains to be done. For stakeholders, policymakers, and scholars interested in how different institutions enable individuals to solve environmental problems, we have illustrated how the framework presented in Chapter 1 helps to organize diagnostic, analytic, and prescriptive capabilities. Using the framework can also aid in the accumulation of knowledge from empirical studies and in the assessment of past efforts at reforms. Instead of analyzing ecological and human systems with a unique language for each puzzle and each setting, we have attempted to develop and use a much more general language that can be used across disciplines, and even across broad clusters of disciplines.

10.1 Connecting real and model worlds

10.1.1 Using our framework to understand diverse socio-ecological systems

It should be clear by now that despite the extraordinary diversity among particular systems—biological or non-biological renewable resources, informal and relatively simple norms to highly articulated formal governance—universal components are nonetheless embedded in any model of ecosystems linked to one or more human systems. Using this framework, there are many ways to characterize ecosystems. In the models developed in Sections II and III, we have defined each spatial unit by specifying values for the following variables:

1. The ecological carrying capacity of the system.
2. The degree to which external influences are predictable.
3. The regeneration rate (population growth rate) of the resource.
4. The natural mortality rate of the resource units.
5. The transfer rate of resource units (or harvesters) from one.
6. The consumption rate(s) of resource users.

In our models in Section 10.2, the carrying capacity of the resource (1 above) was set to a specified constant level throughout each analysis. In Chapter 3, we explored variations in the degree to which external influences are predictable (2), the regeneration rate (3), the natural transfer rate (5), and the robustness of different harvest rules about consumption rates (6). It is also apparent from Chapter 3 that the importance of different ecological variables differs depending on the settings of the values of the other ecological variables. In Chapter 4, we placed particular emphasis on the effects of variation in the mortality rate (4), and consider an issue seldom raised for marine systems: what is the effect on sustainability if the population is not panmictic, but structured? We find reason for concern. Local populations can be "winked out" while managers at a variety of levels expect that local populations will be replenished because they assume the population is panmictic. In Chapter 5, for irrigation systems, we explored the impact of changes in variables (2), (4), and (6). In Chapter 9, all five variables were varied to explore their impacts on management of the Patuxent watershed. The Patuxent model is the most developed and spatially realistic model in this volume, modeling several ecological variables as a spatially realistic mosaic.

In this diverse array of models, we have found some strong and consistent general relationships. First, as unpredictability increases, the failure rate of resource systems increases, even in systems in which the rules are very well matched to local circumstances (e.g., Chapters 3, 4, 5), reinforcing Levins' (1966) point that variation is a cost in any system (and unpredictable variation is especially costly). All humanly designed systems can fail. The greater the environmental unpredictability faced by managers, the greater the probability of failure. As W. Ross Ashby (1960) established in his "law of requisite variety," any sustainable regulatory system must have as much variety in its response capabilities as variety exists in the environment itself. Second, lack of knowledge by decision makers, like environmental uncertainty, can generate high costs and/or failures (best shown in Chapters 3 and 4). Third, high rates of transfer of materials, resources (Chapters 3, 5, 9), or harvesters (Chapter 4) across ecological or social boundaries pose problems that may require both local and higher-order rules to solve. As we discuss further below, lack of knowledge can lead to scale mismatches. Variation in local units may be substantial, and local resource users may have better specific local ecological knowledge than central decision makers, who may have better scientific knowledge about some of the core relationships in a larger set of systems. On the other hand, small-scale organization may fail to gain economies of scale and be unable to draw on the best scientific information available. Large-scale organizations typically have difficulties gaining access to extensive time and place information and responding rapidly to discrete signals from diverse ecological and social systems.

Of course, ecological knowledge is not the only component in decision making; naive prescriptions for both central and local decision making have been shown to be inadequate. Nesting smaller-scale organization within larger-scale organizations enables stakeholders and policy makers to obtain some of the advantages of both scales of organization.

10.1.2 The analysis of possible worlds: future directions

Throughout this book we have argued that sustainability is enhanced when humans design rules that are "well tailored" to an ecosystem. We have argued that many resource-management failures arise from a mismatch in scale between ecological realities and human rules (Cleveland et al., 1996). As we noted in Chapter 1, large-scale systems are not small-scale systems writ large, nor are small-scale systems mere miniatures of large systems. We suggest that management systems that produce perfectly acceptable outcomes in ecosystems at one level can produce disruptive or destructive results when applied to larger- or smaller-scale systems. For example, when high rates of natural capital transfer exist, but are unrecognized so that only local rules are made, voracious harvesters in one sub-system endanger the sustainability of more than their local area (Chapters 3 and 5). Or, when rules are made assuming uniform conditions and full exchange of natural capital across a large region, when, in fact, resource stocks are not uniform (as in structured metapopulations of oceanic fish stock), managers can inadvertently destroy widespread, but structured, populations (Chapter 4).

In proposing the framework of Chapter 1, we argued for a variety of reasons that it was important to link human decision making systems and ecosystems. A major reason, we claimed, was that resource systems are likely to fail for unrecognized reasons unless we overtly study how human systems and ecosystems are linked or not, and how they interact in more complete ways than traditional (e.g., resource management) models. Studying these relationships allows one to identify missing institutions and scale mismatches between institutions and ecological realities.

10.1.3 Mismatches between ecological conditions and decision making

It appears that many current governance and management systems are at a scale that is either too large or too small for the ecosystems to which they are related, leading to unsustainable policies for these systems (e.g., Ostrom et al., 1999). Perhaps this is not surprising, for the growth of many institutions is driven more by politics than by ecology. Problems can occur when human systems that are developed and sustainable at one scale, or for one ecosystem or for one part of an ecosystem are transferred without adequate modification to other scales and ecosystems or to the whole system (Ostrom, 1999).

An important influence on the models of Chapters 3 through 6 is the transfer rate of resources (and in Chapter 4, also the movement of resource users) among spatial units. In some systems there are resource transfers, in others there are not; in some systems there are human rules about these transfers, in others there are not. When natural capital stocks (or resource users) transfer but no decision making capacity exists—in other words, when appropriate institutions are missing—sustainability is imperiled. Perhaps the best-known problems of missing institutions is that of open access to common-pool resources, modeled in Chapters 3 and 4 (see also Hardin, 1968). When no users can be excluded from a valuable resource, and no rules for harvest exist, resource failure is virtually universal (Ostrom et al., 1999).

Consider a simple scheme for relating the presence or absence of local or higher-level rules to a key flow characteristic of ecosystems—the transfer rate of resource stocks from one ecosystem unit to another. Figure 10.1 represents a simplified three-dimensional array of natural transfer rates of resource stocks, local harvest rules, and higher-level rules. Light-shaded cells represent an appropriate match of rules and conditions, while dark-shaded cells represent conditions in which appropriate rules are missing.[1]

Imagine a fishing village on an isolated lake. The natural transfer of biological units between spatial units would be extremely low or nonexistent because fish do not migrate into or out of the lake, and transfers such as accidental human introductions will be rare. For simplicity's sake, we begin with a case in which there are no higher-level governance units. Such an isolated system is represented by cell L11 in Figure 10.1 (if there are also no local governance rules), or cell L12 (if local rules are present). Without local rules (cell L11), the use of the lake could be sustainable if two conditions are met (1) fish are taken simply for local subsistence consumption, and (2) the population of the village is low enough that fish harvests do not exceed the fish production

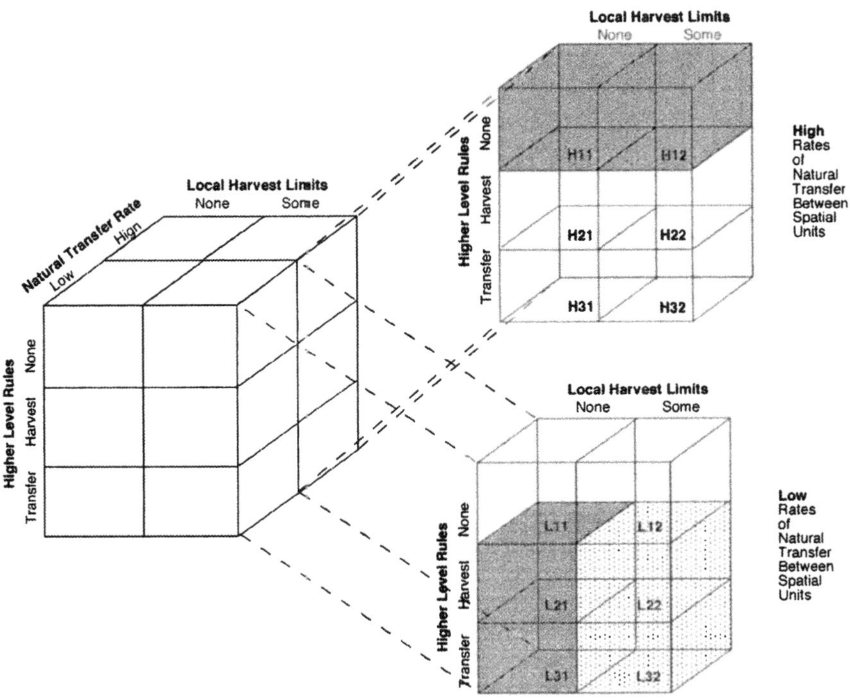

Figure 10.1 A simplified three-dimensional array of natural transfer rates of resource stocks, local harvest rules, and higher level rules

[1] Note, however, that there are some additionally constrained conditions in which the absence of rules may not pose a problem; for example, when human population is low and the resource is used only for subsistence.

capacity of the lake. If the human population grows, however, it is likely either that local users would devise rules to regulate harvest, and thus move from cell L11 to cell L12, or that harvests would exceed sustainable levels and lead to the collapse of the fish population. This general pattern means that local rules are likely to limit either who can fish or how much fish they can harvest—both effectively concern harvest rates.

In summary, cell L11 has neither local nor higher-level rules. Cells L21, L31, H11, H21, and H31 have no local rules; cells L12, H11, and H12 have no higher-level rules. In all of these cases, the problem of missing institutions may lead to unsustainability.

As noted in Chapter 3, the Maine lobster fisheries represent the conditions of cell H12 of Figure 10.1. Although only older, mature (marketable) lobsters migrate, migration creates significant biological transfers when mature lobsters move from inshore waters (where one set of local rules obtains) to offshore waters (where different local rules exist). In this example, there is high biological transfer across governance units, local harvest rules, low or no higher-level control—and indeed human rules and ecological realities do not match. At a larger scale, eliminating restraint for Maine fishermen would simply have meant higher takes in all areas and likely over-exploitation typical of open-access resources. Maine fishermen took their case to the Atlantic States Marine Fisheries Commission (ASMFC)—an interstate body that has authority to govern lobster conservation. The ASMFC agreed with the position of Maine lobstermen and extended the area in which v-notch and oversize protection is enforced to the entire Gulf of Maine. The designated area does not entirely contain the movements of lobsters (due to political compromises within ASMFC and also due to the fact that it is probably impossible to draw a perfect ecological boundary). The improvement brought about by the actions of the ASMFC, however, substantially aligns the scale of human rules and the biologically relevant behavior.

It is easy to imagine how a first approach to scale mismatches like this one could generate the old maxim that "harvest rules should apply over the entire range of the managed stock" to avoid conflicting incentives. But this maxim captures only a small part of the problem. Like most maxims, it originates under particular conditions (here, cell H12 of Figure 10.1); yet it is typically stated quite generally. A moment's reflection will make it obvious that this maxim would result in poor sustainability in a number of other cells (e.g., L21, L31) of Figure 10.1.

The danger is that maxims can lead to management systems, and when maxim-based management systems are applied broadly without investigating the real causal factors, unintended consequences can result. In the Maine lobster fishery, for example, fishermen in some areas have strong collective local incentives to limit trap numbers because high trap numbers in the area lead to congestion and gear conflicts. Because ecological and fishing circumstances vary widely along the coast, a trap limit that is reasonable in one area might be outrageously high or low in another, with unintended local variation in impacts (in this case, the economics of congestion). Efforts to solve the problem at the statewide level have failed for over 20 years because a single trap limit tends to restrain and benefit fishermen in some areas, but imposes costs without benefits in others. In 1995, however, the legislature created a system of seven democratically elected local management councils. Within a year of their initial operation, all seven had enacted trap limits.

In Chapter 1, we suggested that scale mismatch occurs when human decisions are either needed but lacking, or made to the wrong scale. What have we learned?

10.2 Empirical explorations: transfer rates and human rules

We made a broad conceptual distinction between physical systems that have very low natural transfer rates and those with high-transfer rates. Examples of the former include isolated grazing areas, small forests that are separated from one another, lagoon fisheries, and lakes. Examples of high-transfer environments include inshore fisheries that are strongly affected by the movement of fish from ocean and other inshore fisheries, inter-connected groundwater basins, and large-scale river systems. Human activity probably accelerates the rate of transfer of ecosystem components, e.g., all non-point source pollution; CO_2 accumulation and the greenhouse effect; nutrients in the Patuxent example of Chapter 9; and the introduction of non-native species such as Zebra mussels to the Great Lakes.

In Figure 10.1, the stock transfer rate is an ecological "given," representing a constraint for human rules systems. Most human rules systems (if they recognize the importance of transfer rates) make rules related to the migration of people into and out of a local unit (see, for example, Bates, 1976; Zucker, 1986).

But the transfer rate is only one phenomenon affecting human rule-making: natural productivity ("rich" versus "poor" environments), external fluctuations (constant, predictably or unpredictably fluctuating), and stock growth or regeneration rate, for example, also matter. We have not modeled all combinations of these conditions but even the simple general model in Chapter 3 has that capacity. In our description of relevant ecosystems, we consider some examples in which these additional factors vary. As Chapters 3 through 9 suggest, dynamic modeling can be an important path to making tractable analyses of extraordinary complexity.

10.2.1 Low-transfer environments (all "L" cells)

In all "L" cells of Figure 10.1, humans do not face the problems (or opportunities) associated with natural resource transfers in and out. When low-transfer environments are poor, both in endowment and in their capacity to grow, they are difficult to use intensively; it is hard to amass human capital rapidly. So long as the human population using such a resource remained small and they used the resource primarily for subsistence, they could use such a natural resource for long periods of time. If the natural mortality rate were high and external influences highly unpredictable, the environment would probably be too harsh for anything but very limited human use. The Masai, Nuer, Arunta, and !Kung people are probably examples of people who have survived in this broad set of worlds.

What kind of rules might work in such an environment? Let us focus on the subsets of this environment in which external influences are relatively unpredictable versus those that are more predictable. In more unpredictable systems, humans may develop rules that allow them to range over large territories, moving themselves to areas that have the best yield at any moment to substitute for the lack of movement of the resource units. Trans-human societies tend to exist in the more unpredictable ecosystems. Note that trans-humanity also occurs in high-transfer, low-productivity environments, and therefore humans track the resource stock movement. In such cases, large-scale transfer rules would be important. Property rules that give groups the right to harvest from a variety of spatial units increase the "transfers" of resource units to different users across distinct ecosystems (compare the fishermen of Chapter 4). When boundaries among ecosystems become "tight" by human design (such as national boundaries), there may be problems of regulation that did not exist prior to these efforts to stop the movement of people (Duany and Duany, 1999).

In low-productivity, low-transfer environments humans typically take what they can get wherever they are and rules that limit the quantity of harvest in each spatial unit are probably very important. Setting harvesting levels close to MSY in any unpredictable and low-return system might be disastrous, as MSY in bad years would allow the harvesting of stock that is badly needed to regenerate a flow of resource units in the years to come. Easter Island appears to be an example of this type of ecological system—and one that is also characterized by a slow growth rate. The first humans to migrate there found an island that was apparently rich in resources but these had accumulated over the centuries. The fast population growth stimulated by harvesting accumulated natural capital led to a massive depletion of resources before the Easter Islanders were able to impose effective rules on themselves (see Brander and Taylor, 1998; Kirsch, 1997; and Reuveny and Maxwell, 1999). Systems that set harvests based on current resource levels are more sustainable, in theory—but that sustainability is only achievable when the resource estimates are both precise and accurate. When spatial resources are moderately predictable or have a constant external influence, humans might rotate among a smaller set of areas rather than over a very large range. If areas are equally stressed at any one point in time, a trans-human strategy would not make much sense. But a swidden system *would* make sense: people would rotate among units to allow areas to regenerate between uses. Swidden systems appear to have been a successful way of sustainably managing resources when the human population level remains low.

Setting harvesting limits close to the maximum sustained yield (MSY) in any low-transfer system with fluctuations, particularly if fluctuations are unpredictable, could be disastrous. MSY in bad years would allow the harvesting of stock badly needed to regenerate a flow of resource units in the years to come. In Chapter 4, Wilson, et al. models the effects of environmental unpredictabilities on the stability of a "percent stock" harvesting strategy. If measurement were perfect and no uncertainties affected fish recruitment, a very high percentage of stock could be taken. However, unpredictable fluctuations in recruitment can cause the collapse of the fisheries (Chapters 3, 4).

Consider a world that is isolated (low transfer) but has rich natural capital and a high regeneration rate. This is the world of wet tropical forests; resources are rich and regeneration fast. When human population levels are low, there is no need for any harvest limits. This is a well-endowed world that grows back quickly after being harvested. There is more room for error in this world than in the first world, but since it is so rich people may be attracted to it, and without strong harvesting rules, it too may be destroyed. Thus, the presence of harvesting rules may be even more important here than in the harsh worlds where few humans live. How would transfer rates affect outcomes? If there were several connected systems and at least one of them was like this world, it might be a good "refuge," if harvesting is not allowed from this unit but transfer rates are set at a moderate level by a higher level government.

10.2.2 High-transfer environments (all "H" cells)

Resource users in all high-transfer cells in Figure 10.1 face somewhat different problems, and much of the impact in any case will depend on whether the area under consideration is a stock "source" or "sink." Systems with rich natural capital, rapid regeneration rate, and high transfer with other similar systems are the "best case" scenario for human use purposes. When neighboring systems are relatively poor, these are the worlds that provide resource "sources" for other systems—and this relationship underlies many conflicts. Precisely because these worlds are rich and regenerate rapidly, they represent the most desirable resource environments for human use. And, because natural transfer rates are high, resource users in less resource-rich neighboring environments can exploit the natural capital generated in these environments (Chapters 3, 4).

High natural-transfer systems that have slow regeneration rates can be used as "home base" for humans who can (1) utilize resources that have transferred in from other systems, or (2) who can, when current resources are low, move themselves to richer neighboring systems, either seasonally or opportunistically. Even when the regeneration rate is high, these worlds will be used as "home base," and because the regeneration rate, as well as the natural transfer rate, is high, these systems are easier to use in this way.

As we noted above, transfer of both resources and resource users among ecological systems has almost certainly been accelerated in historical time as humans increase their technology. All sorts of problems can arise. Inadvertent export of resource stocks (e.g., zebra mussels, English sparrows) and resource-use byproducts (the Exxon oil spill) become more frequent as we move ourselves, our natural resources, and our manufactured capital around the globe. Deliberate export of stocks (in many natural resource plans) and resource-use by-products (e.g., the "garbage barge" incidents of a few years ago) are also becoming more common. Deliberate export of by-products is widely recognized as creating problems of equity and environmental justice.

10.3 Applications to evolving principles of sustainability

A crucial question facing all who are interested in fostering the sustainability of resource systems is: How can the concepts of analytic models be connected to the real world in ways that foster sustainable resource management? We suggest that models can explore assumptions, be used for experimentation, help individuals or groups with incomplete information, and help groups reach agreement on issues of resource-use policy when conflicts of interest exist.

The difficulties we face in reaching agreement about resource use increase not only when institutions and ecological realities are mismatched in scale (above) but also as the scale of problems increase. We recognize that attempts to achieve globally optimal resource policies in the face of natural and human uncertainty are chimeras. We believe that our best hope lies in including multiple viewpoints in an integrated, adaptive framework structured around a core set of mutually agreed upon principles. A recent workshop in Lisbon, Portugal (Costanza et al., 1998) identified six major principles as a basis for more sustainable management of renewable resources in the future. Reflecting on earlier chapters, we find that these principles are consistent with the findings from models presented in Sections II and III.

10.3.1 Lisbon principle 1: responsibility

Access to environmental resources carries attendant responsibilities to use them in an ecologically sustainable, economically efficient, and socially fair manner. Individual and corporate responsibilities and incentives should be aligned with each other and with social and ecological goals. Chapter 5 illustrates the importance of this principle in one system.

10.3.2 Lisbon principle 2: scale-matching

Ecological problems are rarely confined to a single scale. Decision making on environmental resources should (1) be assigned to institutional level(s) that maximize relevant ecological input, (2) ensure the flow of ecological information between institutional levels, (3) take ownership and actors into account, and (4) internalize costs and benefits. As demonstrated in Chapters 4 and 9, the appropriate scales of governance will be those that have the most relevant information, can respond quickly and efficiently, and are able to integrate across scale boundaries. When the ecological phenomenon occurs at multiple levels, multi-level rules are needed.

10.3.3 Lisbon principle 3: precaution

In the face of uncertainty about potentially irreversible environmental impacts, decisions concerning their use should err on the side of caution. The burden of proof should shift to those whose activities potentially damage the environment. This may be one of the most difficult of the principles to implement. Chapters 3 and 4 explore consequences of both the application and the failure to apply this principle.

10.3.4 Lisbon principle 4: adaptive management

Given that some level of uncertainty always exists in environmental resource management, decision-makers should, to the extent feasible, continuously gather and integrate appropriate ecological, social, and economic information with the goal of adaptive improvement. Chapters 3, 4, 5, and 7 explore the effects of unpredictable environmental variations and the utility of more- and less-adaptive management responses.

10.3.5 Lisbon principle 5: full cost allocation

All of the internal and external costs and benefits, both social and ecological, of alternative decisions concerning the use of environmental resources should be identified and allocated. When appropriate, markets should be adjusted to reflect full costs and allocate them appropriately. We recognize that this is a non-trivial problem but suggest that explicit consideration of this principle may highlight successful paths. Chapter 5 illustrates how cost-benefit analysis is frequently adopted in the early stages of considering a project, but may not include room for the diversity of incentives and actions actually taken by stakeholders once a new infrastructure has been constructed. Chapter 5 also considers the appropriate discount rate to be used in the analysis of projects that may last for more than one century.

10.3.6 Lisbon principle 6: participation

All stakeholders should be engaged in the formulation and implementation of decisions concerning environmental resources. Full stakeholder participation contributes to credible, accepted (and therefore lasting) rules that identify and assign the corresponding responsibilities appropriately. As we show in Chapter 5, when major participants do not contribute to institutional development and maintenance, the likelihood of their achieving sustainable resource systems is substantially reduced.

As are most principles, these are ideals—normative claims about the ways we should proceed. Consider the circumstances in which people are more, versus less, likely to adopt such principles. Some are well known: small, stable groups, perhaps kin, with the ability to detect and punish cheaters, and with developed conflict-resolution mechanisms (e.g., Ostrom, 2000). These involve self-interest and the protection of that interest. A second set of factors that influence people's willingness to invest in the long term concerns information. If users are dependent on the resource, information about practices that clearly degrade the resource mean that the shadow of the future can be long. One source of information, of course, is empirical: we learn from the successes and failures of those around us. A second and complementary resource of information comes from the kinds of models we have explored in earlier chapters.

10.4 Last reflections

Throughout, we have argued that sustainability depends on understanding the ways humans and their institutions interact with ecological systems. Both information and consideration of actors' self-interests are necessary, but not sufficient, inputs to sustainable management (Wilson et al., 1994; Wilson, 1997). The growing litany of ecological problems we see daily in the news typically involves issues of increasing human populations, increasingly intrusive production mechanisms, habitat and species destruction—all symptoms of human–ecosystem interactions gone awry. Using the integrated framework developed in Chapter 1 to link human-ecosystem interactions, we have begun explorations using dynamic modeling. We think there is profit in such an approach. Such models can be used analytically to generate information complementary to empirical data, to aid in thoughtful decision-making, and to help build consensus in contentious situations.

We suggest that when the scales of ecological phenomena and human rules for their governance are appropriately matched, governance systems can be responsive and appropriate. Notice that we do not claim that humanly designed governance systems will be optimal. Designing optimal governance systems to cope with multiple, complex ecological systems is far more difficult and unlikely than our current textbooks on resource management imply. We do, however, believe that individuals can design and operate under appropriate rules that allow sufficient information to be generated over time to enable participants to learn from their past mistakes and continuously improve the systems they design. Whether a local or larger-scale rule is appropriate and adaptive depends on the scale at which the relevant ecological interactions take place. Making this problem of scale-matching explicit is a primary goal of our enterprise here. Consider, for example, if resource stocks are spatially isolated and no transfer of natural capital occurs, what use are higher-order rules? In fact, if the isolated resource systems are ecologically different in important ways, uniform, higher-level rules can be costly in more ways than one. On the other hand, consider the Maine lobster example above and in Chapter 3, in which it is difficult to apply conserving local regulations throughout a large area. Slowly, more and more local units are applying these conservation regulations so that eventually a much larger portion of the northeastern coast of the United States may be covered, but it is a slow and difficult process.

Thus, since there is so much to be done, we hope that other scholars will find the frameworks, concepts, and tools we have developed in this book to be useful in their own efforts to examine these important questions. We will gladly make our models available to anyone who would like to use them, and encourage a wider discussion across disciplines of these important questions.

References

Abbott, M.B., et al., 1986. An introduction to the European Hydrological System—Systeme Hydrologique Europeen, SHE 2: Structure of a physically-based, distributed modelling system. *J. Hydrol.*, 87: 61–77.

Alexander, R.D., 1987. *The Biology of Moral Systems.* Aldine De Gruyter, New York.

Allen, T.F.H., Starr, T.B., 1982. *Hierarchy.* University of Chicago Press, Chicago.

Ames, T., 1996. *Cod and Haddock Spawning Grounds in the Gulf of Maine.* Island Institute, Rockland, ME.

Anderson, L.G., 1986. *The Economics of Fisheries Management.* Johns Hopkins University Press, Baltimore, MD.

AQUA TERRA, 1994. *Patuxent River Basin Watershed Model.* Report for MDE and USGS: 94.

Arneado, Coullet, Peyraud, and Tresser. 1982. Strange attractors in Volterra equations for species in competition. *J. Math Biol.*, 14: 153–157.

Arnold, J.E.M., 1998. *Managing Forests as Common Property.* FAO Forestry Paper, 136, Rome.

Arrow, K., 1951. *Social Choice and Individual Values,* 2nd ed., Wiley, New York.

Ashby, W.R., 1960. *Design for a Brain: The Origin of Adaptive Behavior,* 2nd ed., Wiley, New York.

Atran, S., 1993. Itza Maya tropical agro-forestry. *Curr. Anthropol.*, 34: 633–689.

Bahr, L.M., Costanza, R., Day, J.W., Jr., Bayley, S.E., Neill, S.E., Leibowitz, S.G., and Fraci, J., 1983. Ecological characterization of the Mississippi Deltaic Plain Region: a narrative with management recommendations. U.S. Fish and Wildlife Service, Division of Biological Services, Washington, D.C. FWS/OBS-82/69.

Bandel, V.A., and Heger, E.A., 1994. *MASCAP, Maryland's Agronomic Soil Capability Assessment Program.* Cooperative Extension Service, University of Maryland, College Park.

Bates, R.H., 1976. *Rural Responses to Industrialization: A Study of Village Zambia.* Yale University Press, New Haven, CT.

Beasley, D.B., and Huggins, L.F., 1980. *ANSWERS (Areal Nonpoint Source Watershed Environment Response Simulation), User's Manual.* Purdue University, West Lafayette.

Bell, K., and Bockstael, N., 2000. Applying the generalized-moments estimation approach to spatial problems involving micro-level data. *Rev. Econ. Stat.*, 82: 72–82.

Berkes, F., Ed., 1989. *Common Property Resources: Ecology and Community-Based Sustainable Development.* Belhaven Press, London.

Berkes, F., and Folke, C., Eds., 1998. *Linking Social and Ecological Systems: Management Practices and Social Mechanisms for Building Resilience.* Cambridge University Press, Cambridge.

Beven, K.J., 1993. Prophecy, reality and uncertainty in distributed hydrological modelling. *Adv. Water Res.*, 16: 41–51.

Biggs, R.B., 1981. Freshwater inflow to estuaries, short and long-term perspectives. In *Proceedings of the National Symposium on Freshwater Inflow to Estuaries*, R.D. Cross and D.L. Williams, Eds., Coastal Ecosystems Project, O.B.S., USFWS. Washington, D.C., 2: 305–321.

Bishop, R.C., 1993. Economic efficiency, sustainability, and biodiversity. *Ambio.*, 22: 69–73.

Bockstael, N.E., 1996. Economics and ecological modeling: the importance of a spatial perspective. *Am. J. Agric. Econ.*, 78: 1168–1180.

Bockstael, N., and Bell, K., 1997. Land use patterns and water quality: the effect of differential land management controls. In *International Water and Resource Economics Consortium: Conflict and Cooperation on Trans-Boundary Water Resources.* R. Just and S. Netanyahu, Eds.

Bockstael, N., Costanza, R., Strand, I., Boynton, W., Bell, K., and Wainger, L., 1995. Ecological economic modeling and valuation of ecosystems. *Ecol. Econ.*, 14: 143–159.

Bond, P. and Goldblatt, P., 1984. Plants of the Cape flora. *J. S. Afr. Bot. Suppl.*, 13: 1–455.

Boyer, M.C., 1964. *Streamflow Measurements. Handbook of Applied Hydrology.* V.T. Chow, Ed. McGraw-Hill, New York, 15.1–15.42.

Brander, J., and Taylor, M.S., 1998. The simple economics of Easter Island: a Ricardo-Malthus Model of renewable resource use. *Am. Econ. Rev.*, 88: 119–138.

Briscoe, J., 1979. *The Political Economy of Energy Use in Rural Bangladesh.* Mimeo.

Bromley, D., Feeny, D., McKean, M., Peters, P., Oakerson, R., and Thomson, J., Eds., 1992. *Making the Commons Work.* Institute of Contemporary Studies, San Francisco.

Brown, B., 1995. In *Timber Country: Working People's Stories of Environmental Conflict and Urban Flight.* Temple University Press, Philadelphia.

Brown, G.M., and Roughgarden, J., 1992. An ecological economy: notes on harvest and growth. *Beijer Discussion Paper Series No. 12.* Beijer International Institute of Ecological Economics, Stockholm, Sweden.

Brown, G.M., and Swierzbinski, J., 1985. Endangered species, genetic capital and cost-reducing R&D. pp. 111–127 In *Economics of Ecosystems Management.* D.O. Hall, N. Myers, and N.S. Margaris, Eds., Dr. W. Junk Publishers, Dordrecht.

Brush, G.S., and Davis, F.W., 1984. Stratigraphic evidence of human disturbance in an estuary. *Q. Res.* 22: 91–108.

Brush, G.S., Lenk, C., and Smith, J., 1980. The natural forests of Maryland: an explanation of the vegetation map of Maryland. *Ecol. Monogr.*, 50: 77–92.

Buchanan, J.T., Acevedo, W., Richards, L.R., 1998. Preliminary assessment of land use change and its effects on the Chesapeake Bay. Ames Research Center. USGS Science for a Changing World Poster.

Carney, J., and Watts, M., 1990. Manufacturing dissent—work, gender and the politics of meaning in a peasant society. *Africa*, 60: 207–241.

Caswell, H., 1989. *Matrix Population Models.* Sinauer Associates, Sunderland, MA.

Cernea, M., 1985. Putting people first: sociological variables. In *Rural Development.* Oxford University Press, New York.

Checkland, P., 1989. Soft systems methodology. In *Rational Analysis for a Problematic World*, J. Rosenhead, Ed., John Wiley & Sons, Chichester.

Cheng, H.T., and Townsend, R.E., 1993. Potential impact of seasonal closures in the U.S. lobster fishery. *Mar. Resour. Econ.* 8: 101–117.

Clark, C., 1976. *Mathematical Bioeconomics, The Optimal Management of Renewable Resources.* John Wiley & Sons, New York.

Clark, C., 1990. *Mathematical Bioeconomics: The Optimal Management of Renewable Resources.* Wiley Interscience, New York.

Clark, C.W., and Munro, G.R., 1975. The economics of fishing and modern capital theory: a simplified approach. *J. Environ. Econ. Manage.,* 2: 92–106.

Cleveland, C.J., 1991. Natural resource scarcity and economic growth revisited: economics and biophysical perspectives. In *Ecological Economics: The Science and Management of Sustainability,* R. Costanza, Ed., Columbia University Press, New York.

Cleveland, C.J., Neill, C., and Day, J.W., Jr., 1981. The impact of artificial canals on land loss in the Barataria Basin, Louisiana. pp. 425–434. In *Energy and Ecological Modeling,* W.J. Mitsch, R.W. Bosserman, and J.M. Klopatek, Eds., Elsevier, Amsterdam.

Cleveland, C., Costanza, R., Eggertsson, T., Fortmann, L., Low, B., McKean, M., Ostrom, E., Wilson, J., and Young, O., 1996. A framework for modeling the linkages between ecosystems and human systems. *Beijer Discussion Paper Series No. 76.* Beijer International Institute of Ecological Economics, Stockholm, Sweden.

Coleman, J., 1988. Social capital in the creation of human capital. *Am. J. Soc.,* 94: 95–120.

Colwell, R.K., 1974. Predictability, constancy, and contingency of periodic phenomena. *Ecology,* 55: 1148–1153.

Conrad, J.M., and Clark, C.W., 1987. *Natural Resource Economics: Notes and Problems.* Cambridge University Press, Cambridge.

Cooper, S.R., 1995. Chesapeake Bay watershed historical land use impact on water quality and diatom communities. *Ecol. Appl.,* 5: 703–723.

Correll, D.L., 1983. N and P in Soils and Runoff of Three Coastal Plain Land Uses. Smithsonian Institution, Chesapeake Bay Center for Environmental Studies Special Pub. No. 23.

Correll, D.L., Jordan, T.J., and Weller, D.E., 1992. Nutrient flux in a landscape: effects of coastal land use and terrestrial community mosaic on nutrient transport to coastal waters. *Estuaries,* 15: 431–442.

Costanza, R., 1987. Simulation modeling on the Macintosh using STELLA. *BioScience,* 37: 129–132.

Costanza, R., Ed., 1991. *Ecological Economics: The Science and Management of Sustainability.* Columbia University Press, New York.

Costanza, R., and Cornwell, L., 1992. The 4P approach to dealing with scientific uncertainty. *Environment* 34: 12–20. (Also reprinted in *Taking Sides: Clashing Views on Controversial Issues in Science, Technology, and Society,* T. Easton, Ed., 1994. Dushkin Press, Guilford; and as pp. 417–426 in *Frontiers in Ecological Economics: Transdisciplinary Essays of Robert Costanza,* Costanza, R., 1997. Edward Elgar, Cheltenham).

Costanza, R., and Greer, J., 1995. The Chesapeake Bay and its watershed: a model for sustainable ecosystem management? pp. 169–213 In *Barriers and Bridges to the Renewal of Ecosystems and Institutions,* L.H. Gunderson, C.S. Holling, and S. Light Eds., Columbia University Press, New York.

Costanza, R., and Maxwell, T., 1991. Spatial ecosystem modeling using parallel processors. *Ecol. Model.,* 58: 159–183.

Costanza, R., and Maxwell, T., 1993. Resolution and predictability: an approach to the scaling problem. *Land. Ecol.,* 9: 47–57.

Costanza, R., and Maxwell, T., 1994. Resolution and predictability: an approach to the scaling problem. *Land. Ecol.*, 9: 47–57.

Costanza, R., and Patten, B.C., 1995. Defining and predicting sustainability. *Ecol. Econ.*, 15: 193–196.

Costanza, R., and Sklar, F.H., 1985. Articulation, accuracy, and effectiveness of mathematical models: a review of freshwater wetland applications. *Ecol. Model.*, 27: 45–68.

Costanza, R., Neill, C., Leibowitz, S.G., Fruci, J.R., Bahr, L.M., Jr., and Day, J.W., Jr., 1983. Ecological models of the Mississippi deltaic plain region: data collection and presentation. U.S. Fish and Wildlife Service, Division of Biological Services, Washington, D.C. FWS/OBS-82/68.

Costanza, R., Sklar, F.H., White, M.L., and Day, J.W., Jr., 1988. A dynamic spatial simulation model of land loss and marsh succession in coastal Louisiana, pp. 99–114 In *Wetland Modelling,* W.J. Mitsch, M. Straskraba, and S.E. Jørgensen, Eds., Elsevier, Amsterdam.

Costanza, R., Sklar, F.H., and White, M.L., 1990. Modeling coastal landscape dynamics. *BioScience*, 40: 91–107.

Costanza, R., Norton, B., and Haskell, B.J., Eds., 1992. *Ecosystem Health: New Goals for Environmental Management.* Island Press, Washington, D.C.

Costanza, R., Wainger, L., Folke, C., and Mäler, K.-G., 1993. Modeling complex ecological economic systems: toward an evolutionary, dynamic understanding of people and nature. *BioScience*, 43: 545–555.

Costanza, R., d'Arge, R., de Groot, R., Farber, S., Grasso, M., Hannon, B., Naeem, S., Limburg, K., Paruelo, J., O'Neill, R.V., Raskin, R., Sutton, P., and van den Belt, M., 1997. The value of the world's ecosystem services and natural capital. *Nature*, 387: 253–260.

Costanza, R., Andrade, F., Antunes, P., van den Belt, M., Boersma, D., Boesch, D.F., Catarino, F., Hanna, S., Limburg, K., Low, B., Molitor, M., Pereira, G., Rayner, S., Santos, R., Wilson, J., and Young, M., 1998. Principles for sustainable governance of the oceans. *Science*, 281: 198–199.

Cowling, R.M., Ed., 1992. *The Ecology of Fynbos. Nutrients, Fire and Diversity.* Oxford University Press, Cape Town, South Africa.

Cowling, R.M., Holmes, P.M., and Rebelo, A.G., 1992. Plant diversity and endemism. In *The Ecology of Fynbos. Nutrients, Fire and Diversity.* R.M. Cowling, Ed., pp. 62–112. Oxford University Press, Cape Town, South Africa.

Craig, N.J., Turner, R.E., and Day, J.W., Jr., 1979. Land loss in coastal Louisiana. *Environ. Manage.*, 3: 133–144.

Crawford, S. and Ostrom, E., 1995. The grammar of rules. *Am. Pol. Sci. Rev.*, 89: 582–600.

Crum, J. R., Pierce, F.J., Vitosh, M.L., and Knezek, B.D., 1990. Potential Nitrogen Contamination of Groundwater as Affected by Soil, Water, and Land Use Relationships. Michigan State University, Dept. of Crop and Soil Sciences. Tech. Rpt. No. 89-G1569–04.

Dawkins, R., 1982. *The Extended Phenotype: The Gene as the Unit of Selection.* Oxford University Press, Oxford.

Debreu, G., 1974. Excess demand functions. *J. Math. Econ.*, 1: 15–23.

Deegan, L.A., Kennedy, H.M., and Neill, C., 1984. Natural factors and human modifications contributing to marsh loss in Louisiana's Mississippi River deltaic plain. *Environ. Manage.*, 8: 519–528.

Detenbeck, N.E., DeVore, P.W., Niemi, G.J., and Lima, A., 1992. Recovery of temperate-stream fish communities from disturbance: a review of case studies and synthesis of theory. *Environ. Manage.* 16: 33–53.

Donigian, A.S., Imhoff, J.C., Bicknell, B.R., and Kittle, J.L., 1984. Application Guide for Hydrological Simulation Program—FORTRAN (HSPF). Athens, GA, U.S. EPA, Environmental Research Laboratory.

Doorenbos, J. and Kassam, A.H., 1979. FAO Irrigation and Drainage Paper No. 33. FAO, Rome.

Duany, W. and Duany, J., 1999. Genesis of the crisis in the Sudan. In *White Nile, Black Blood,* J. Spaulding and S. Beswick, Red Seas Press, Lawrenceville, N.J. pp. 167–182.

Duchin, F., 1992. Industrial input-output analysis: implications for industrial ecology. *Proc. Natl. Acad. Sci. U.S.A.* 89: 851–855.

Dugatkin, L.A., 1997. *Cooperation among Animals: An Evolutionary Perspective.* Oxford University Press, Oxford.

Dugatkin, L.A. and Reeve, H.K., Eds., 1998. *Game Theory and Animal Behavior.* Oxford University Press, Oxford.

Dunne, T. and Leopold, L.B., 1978. *Water in Environmental Planning.* W.H. Freeman and Company, New York.

Eggertsson, T., 1990. *Economic Behavior and Institutions.* Cambridge University Press, New York.

El Serafy, S., 1989. The proper calculation of income from depletable natural resources. In *Environmental Accounting for Sustainable Development. A UNEP—World Bank Symposium,* Y.J. Ahmad, S. El Serafy, and E. Lutz, Eds., The World Bank, Washington, D.C., pp. 10–18.

FAO/UNESCO, 1973. *Irrigation, Drainage and Salinity: An International Source Book.* Hutchinson and Company Ltd., London, England.

Finlayson, A.C., and McCay, M.J., 1998. Crossing the threshold of ecosystem resilience: the commercial extinction of northern cod. In *Linking Social and Ecological Systems: Management Practices and Social Mechanisms for Building Resilience,* F. Berkes and C. Folke, Eds. Cambridge University Press, New York.

Fitz, H.C., Costanza, R., and Reyes, E., 1993. The Everglades Landscape Model (ELM): Summary Report of Task 2, Model Development. Report to the South Florida Water Management District, Everglades Research Division.

Fitz, H.C., Voinov, A., and Costanza, R., 1995. The Everglades Landscape Model: Calibration Analysis Report. Report to the Everglades Systems Research Division, South Florida Water Management District.

Fitz, H.C., DeBellevue, E., Costanza, R., Boumans, R., Maxwell, T., Wainger, L., and Sklar, F., 1996. Development of a general ecosystem model for a range of scales and ecosystems. *Ecol. Model.,* 88: 263–295.

Flather, C.H. and Sauer, J.R., 1996. Using landscape ecology to test hypotheses about large-scale abundance patterns in migratory birds. *Ecology,* 77: 28–35.

Folke, C., Hammer, M., Costanza, R., and Jansson, A.M., 1994. Investing in natural capital: why, what, and how? pp. 1–20. In *Investing in Natural Capital: The Ecological Economics Approach to Sustainability,* A.M. Jansson, M. Hammer, C. Folke, and R. Costanza, Eds., Island Press, Washington, D.C.

Forgarty, M., Hilborn, R., and Gunderson, D., 1997. Chaos and parametric management. *Mar. Pol.,* 21: 187–194.

Gagliano, S.M., Meyer-Arendt, K.J., and Wicker, K.M., 1981. Land loss in the Mississippi River deltaic plain. *Transactions of the Gulf Coast Association of Geological Societies,* 31: 285–300.

Gardner, R.H., Castro, M.S., Morgan, R.P., and Seagle, S.W., 1996. Nitrogen dynamics in forested lands of the Chesapeake Basin. *STAC Perspectives on the Chesapeake Bay. No. 151,* Chesapeake Research Consortium, Inc., Edgewater, MD.

Gause, G.F., 1935. *La Theorie Mathematique de la Lutte Pour la Vie.* Herman, Paris.

Geoghegan, J., Wainger, L., and Bockstael, N., 1997. Spatial landscape indices in a hedonic framework: an ecological economics analysis using GIS. *Ecol. Econ.,* 22: 251–264.

Gholz, H.L., Linder, S., and McMurtie, R.E., 1994. *Environmental Restraints on the Structure and Productivity of Pine Forest Ecosystems: A Comparative Analysis.* Munksgaard, Copenhagen.

Gibson, C., McKean, M., and Ostrom, E., 2000. *People and Forests: Communities, Forests, and Institutions.* MIT Press, Cambridge, MA.

Gilpin, M., 1996. Metapopulations and wildlife conservation: approaches to modeling spatial structure. In *Metapopulations and Wildlife Conservation,* D.R. McCullough, Ed., Island Press, Washington, D.C., pp. 11–27.

Gordon, H.S., 1954. Economic theory of a common property resource: the fishery. *J. Pol. Econ.,* 62: 124–142.

Grafen, A., 1991. Modeling in behavioural ecology. Chapter 1, In *Behavioural Ecology: An Evolutionary Approach,* 3rd ed., J. R. Krebs and N. B. Davies, Eds., Blackwell Scientific, Oxford.

Granger, C.W.J., 1969. Investigating causal relations by econometric models and cross-spectral methods, *Econometrica,* 37: 424–438.

Granger, C.W.J., 1993. What are we learning about the long run? *Econ. J.,* 103: 307–317.

Groffman, P.M., and Likens, G.E., Eds., 1994. *Integrated Regional Models: Interactions between Humans and Their Environment.* Chapman & Hall, New York.

Gulland, J.A., 1977. The analysis of data and development of models. In *Fish Population Dynamics,* J.A. Gulland, Ed., Wiley, New York, pp. 67–95.

Gunderson, L., Holling, C.S., and Light, S., Eds., 1995. *Barriers and Bridges to the Renewal of Ecosystems and Institutions.* Columbia University Press, New York.

Gustafson, E.J., and Gardner, R.H., 1996. The effect of landscape heterogeneity on the probability of patch colonization. *Ecology,* 77: 94–107.

Haith, D.A., Tubbs, L.J., and Pickering, N.B., 1984. *Simulation of Pollution by Soil Erosion and Soil Nutrient Loss.* Pudoc, Wageningen.

Hall, C., 1988. An assessment of several of the historically most influential theoretical models used in ecology and of the data provided in their support. *Ecol. Model.,* 43: 5–31.

Hall, C.A.S. and Hall, M.H.P., 1993. The efficiency of land and energy use in tropical economies and agriculture. *Agric. Ecosyst. Environ.,* 46: 1–30.

Hamilton, W.D., 1964. The genetical evolution of social behaviour I, II. *J. Theor. Biol.,* 7: 1–52.

Hannon, B., and Joiris, C., 1987. A seasonal analysis of the southern North Sea ecosystem. *Ecology,* 70: 1916–1934.

Hannon, B., and Ruth, M., 1994. *Dynamic Modeling.* Springer-Verlag, New York.

Hardin, G., 1968. The tragedy of the common. *Science,* 162: 1243–1248.

Hawkins, C.P., Gottschalk, L.J., and Brown, S.S., 1988. Densities and habitat of tailed frog tadpoles in small streams near Mt. St. Helens following the 1980 eruption. *J. N. Am. Benthol. Soc.,* 7(3): 246–252.

Heilbruner, R.L., 1974. *An Inquiry into the Human Prospect.* Norton, New York.

Higgins, S.I., Turpie, J. K., Costanza, R., Cowling, R.M., Le Maitre, D.C., Marais, C., and Midgley, G.F., 1997. An ecological economic simulation model of mountain fynbos ecosystems: dynamics, valuation and management. *Ecol. Econ.,* 22: 155–169.

Hilborn, R. and Walters, C.J., 1992. *Quantitative Fisheries Stock Assessment: Choice, Dynamics, and Uncertainty.* Chapman & Hall, New York.

Hillel, D., 1987. The Efficient Use of Water in Irrigation: Principles and Practices for Improving Irrigation in Arid and Semiarid Regions. World Bank Technical Paper No. 64. World Bank, Washington, D.C., pp. 107.

Hobbs, R.J., Richardson, D.M., and Davis, G.W., 1995. Mediterranean-type ecosystems: opportunities and constraints for studying the function of biodiversity. In *Mediterranean-Type Ecosystems. The Function of Biodiversity,* G.W. Davis and D.M. Richardson, Eds., Springer, Berlin, pp. 1–42.

Hofmann, E.E., 1991. How do we generalize coastal models to global scale? In *Ocean Margin Processes in Global Change,* R.F.C. Mantoura, J.M. Martin, and R. Wollast, Eds., John Wiley & Sons, New York, pp. 401–417.

Hogarth, R., 1987. *Judgment and Choice.* John Wiley & Sons, Chichester, England.

Holling, C.S., 1964. The analysis of complex population processes. *Can. Entomol.,* 96: 335–347.

Holling, C.S., Ed., 1978. *Adaptive Environmental Assessment and Management.* Wiley, London.

Holling, C.S., 1987. Simplifying the complex: the paradigms of ecological function and structure. *Eur. J. Oper. Res.,* 30: 139–146.

Hutchins, J. A., and Meyers, R.A., 1995. The biological collapse of Atlantic cod off Newfoundland and Labrador. In *The North Atlantic Fisheries: Successes, Failures, and Challenges,* R. Arnason and L. Felt, Eds., The Institute of Island Studies, Charlottetown, PEI, pp. 37–93.

Jacquez, J. A., and Simon, C.P., 2000. Qualitative Theory of Compartmental Systems with Lags. University of Michigan manuscript.

Jansson, A., Hammer, M., Folke, C., and Costanza, R., Eds., 1994. *Investing in Natural Capital: The Ecological Economics Approach to Sustainability.* Island Press, Washington, D.C.

Johnson, D.W., and Lindberg, S.E., 1992. *Atmospheric Deposition and Forest Nutrient Cycling: A Synthesis of the Integrated Forest Study.* Springer-Verlag, New York.

Jones, J., 1996. USGS, Reston. NDVI data for Patuxent area, personal communication.

Kahnemann, D., and Tversky, A., 1974. Judgment under uncertainty. *Science,* 185: 1124–1131.

Kahnemann, D., Slovic, P., and Tversky, A., 1982. *Judgment Under Uncertainty: Heuristics and Biasis,* Cambridge University Press, Cambridge.

Kaitala, V., and Pohjola, M., 1988. Optimal recovery of a shared resource stock: a differential game model with efficient memory equilibria. *Nat. Res. Model.,* 3: 91–119.

Kemp, W. M., Boynton, W.R., Twilley, R.R., Stevenson, J.C., and Means, J.C., 1983. The decline of submerged vascular plants in Upper Chesapeake Bay: summary of results concerning possible causes. *Mar. Tech. Soc. J.,* 17: 78–89.

Kemp, W. M., Boynton, W.R., Cunningham, J.J., Stevenson, J.C., Jones, T.W., and Means, J.C., 1985. Effects of atrazine and linuron on photosynthesis and growth of the macrophytes, *Potamogeton perfoliatus* L. and *Myriophyllum spicatum* L. in an estuarine environment. *Mar. Envir. Res.,* 16: 255–280.

Keohane, R.O., and Ostrom, E., 1995. *Local Commons and Global Interdependence: Heterogeneity and Cooperation in Two Domains*. Sage, London.
Keynes, J.M., 1936. *General Theory of Employment, Interest and Money*. Harcourt Brace, London.
Khan, H., and Brush, G.S., 1994. Nutrient and metal accumulation in a freshwater tidal marsh. *Estuaries*, 17: 345–360.
Kirch, P.V., 1997. Micocosmic history: island perspectives on "global" change. *Am. Anthropol.*, 99: 30–42.
Klein, L.R., 1971. Forecasting and policy evaluation using large-scale econometric models: the state of the art. In *Frontiers of Quantitative Economics*, M.D. Intriligator, Ed., North-Holland Publishing Co., Amsterdam.
Kraemer, K.L., and King, J.L., 1988. Computer-based systems for cooperative work and group decisionmaking. *ACM Comput. Surv.*, 20: 115–146.
Krysanova, V.P., Gerten, D., Klocking, B., and Becker, A., 1999. Factors affecting nitrogen export from diffuse sources: a modeling study in the Elbe basin. In *Impact of Land-Use Change on Nutrient Loads from Diffuse Sources*, L. Heathwaite, Ed., IAHS Publication 257: 201–212.
Lam, W.F., 1998. *Governing Irrigation Systems in Nepal—Institutions, Infrastructure, and Collective Action*. ICS Press, Oakland, CA.
Lee, K.N., 1993. Greed, scale mismatch, and learning. *Ecol. Appl.*, 3: 560–564.
Leibowitz, S., 1989. The Pattern and Process of Land Loss in Coastal Louisiana: A Landscape Ecological Analysis. Ph.D. dissertation, Louisiana State University, Baton Rouge, LA.
Levins, R., 1966. The strategy of model building in population biology. *Am. Sci.*, 54: 421–431.
Lichtenberg, E., and Shapiro, L.K., 1997. Agriculture and nitrate concentrations in Maryland community water system wells. *J. Envir. Qual.*, 26: 145–153.
Lines, M., 1989. Environmental noise and nonlinear models: a simple macroeconomic example. *Econ. Notes*, 19: 376–394.
Lines, M., 1990. Stochastic stability considerations: a nonlinear example. *Int. Rev. Econ. Bus.*, 37: 219–233.
Low, B.S., 1994. Human sex differences in behavioral ecological perspective. *Anal. Krit.*, 16: 38–67.
Low, B.S., 2000. *Why Sex Matters*. Princeton University Press, Princeton, NJ.
Low, B.S., and Berlin, J.A., 1984. Natural selection and the management of rangelands. In *Developing Strategies in Rangeland Management*, Gardner et al., Westview Press, Boulder, CO.
Low, B.S. and Heinen, J.T., 1993. Population, resources, and environment: implications of human behavioral ecology for conservation. *Pop. Environ.*, 15: 7–41.
Lucas, R.E., 1975. An equilibrium model of the business cycle. *J. Pol. Econ.*, 83: 1113–1145.
Ludwig, D., Hilborn, R., and Walters, C., 1993. Uncertainty, resource exploitation, and conservation: lessons from history. *Science*, 260: 17–36.
Lynch, J.A., and Corbett, E.S., 1990. Evaluation of best management practices for controlling nonpoint pollution from silviculture operations. *Water. Res. Bull.* 26: 41–52.
Lyneis, J.M., 1980. *Corporate Planning and Policy Design: A System Dynamics Approach*. Pugh-Roberts Associates, Cambridge, MA.
Magette, W.L., Weismiller, R.A., Lessley, B.V., Wood, J.D., and Miller, C.F., 1990. Demonstrating agricultural best management practice implementation and impacts on a commercial farm. *ASAE-Appl. Eng. Agric.*, 6: 45–52.

Mäler, K.G., 1991. National accounts and environmental resources. *Environ. Res. Econ.*, 1: 1–15.

Maryland Office of Planning and Maryland Department of the Environment, 1993. Nonpoint Source Assessment and Accounting System. Final Report for FFY '91 Section 319 Grant.

Maxwell, T., and Costanza, R., 1995. Distributed modular spatial ecosystem modeling. *Int. J. Comp. Sim. Spec. Issue Adv. Simulation Methodol.*, 5: 247–262.

May, R., 1974. Biological populations with nonoverlapping generations: stable points, stable cycles, and chaos. *Science.*

McCay, B.J., and Acheson, J.M., 1987. *The Question of the Commons: The Culture and Ecology of Communal Resources.* University of Arizona Press, Tuscon, AZ.

Morecroft, J.D.W., 1994. Executive knowledge, models, and learning. In *Modeling for Learning Organizations,* J.D.W. Morecroft and J.D. Sterman, Eds., Productivity Press, Portland, OR, pp. 3–28.

Morecroft, J.D.W., and van der Heijden, K.A.J.M., 1994. Modeling the oil producers: capturing oil industry knowledge in a behavioral simulation model. In *Modeling for Learning Organizations,* J.D.W. Morecroft and J.D. Sterman, Eds., Productivity Press, Portland, OR, pp. 147–174.

Morecroft, J.D.W., Lane, D.C., and Viita, P.S., 1991. Modeling growth strategy in a biotechnology startup firm. *Sys. Dyn. Rev.*, 7: 93–116.

Mountford, K., 1997. Past is prologue: history lessons for the future. *Bay J., Alliance for the Chesapeake Bay.* 7(1), available at http://web.gmu.edu/bios/bay/journal/97–03/bay.htm.

Murphy, G.I., 1966. Population biology of the Pacific salmon (*Sardinops caerulea*). *Proc. Calif. Acad. Sci.*, 34: 1–84.

Myers, N., 1990. The biodiversity challenge: expanded hot-spots analysis. *Environmentalist,* 10: 243–255.

Nhira, C., 1994. Indigenous Woodland Usage and Management in Kanyati Communal Area, Zimbabwe with Special Reference to Common Property Issues. Ph.D. dissertation, Center for Applied Social Sciences, University of Zimbabwe, Harare, Zimbabwe.

Novotny, V., and Olem, H., 1994. *Water Quality. Prevention, Identification, and Management of Diffuse Pollution.* Van Nostrand Reinhold, New York.

O'Neill, R.V., and Rust, B., 1979. Aggregation error in ecological models. *Ecol. Model.*, 7: 91–105.

O'Neill, R.V., DeAngelis, D.L., Waide, J.B., and Allen, T.F.H., 1986. *A Hierarchical Concept of Ecosystems.* Princeton University Press, Princeton, NJ.

O'Neill, R.V., Krummel, J.R., Gardner, R.H., Sugihara, G., Jackson, B., DeAngelis, D.L., Milne, B.T., Turner, M.G., Zygmunt, B., Christensen, S.W., Dale, V.H., and Graham, R.L., 1988. Indices of landscape pattern. *Landscape Ecol.*, 1: 153–162.

O'Neill, R.V., Johnson, A.R., and King, A.W., 1989. A hierarchical framework for the analysis of scale. *Landscape Ecol.*, 3: 193–205.

Ophuls, W., 1973. Leviathon or oblivion. In *Toward a Steady State Economy,* H.E. Daly, Ed., Freeman, San Francisco, pp. 215–230.

Oreskes, N., Shrader-Frechette, K., and Beliz, K., 1994. Verification, validation and confirmation of numerical models in the earth sciences. *Science,* 263: 641–646.

Orth, R.J., Nowak, J.F., Anderson, G.F., Wilcox, D.J., Whiting, J.R., and Nagey, L.S., 1995. Distribution of submerged aquatic vegetation in the Chesapeake Bay and tributaries and Chincoteague Bay—1994. Final Report, U.S. EPA.

Oster, G., 1996. *Madonna.* http://nature.berkeley.edu/~goster/madonna.html.
Ostrom, E., 1990. *Governing the Commons.* Cambridge University Press, Cambridge.
Ostrom, E., 1999. Social capital: a fad or a fundamental concept? In *Social Capital: A Multifaceted Perspective,* P. Dasgupta and I. Serageldin, Eds., The World Bank, Washington, D.C., pp. 172–214.
Ostrom, E., 2000. Reformulating the commons. In *The Commons Revisited: An American Perspective,* J. Burger, R. Norgaard, E. Ostrom, D. Policansky, and B. Goldstein, Eds., Island Press, Washington, DC.
Ostrom, E., Gardner, R., and Walker, J., 1994. *Rules, Games, and Common-Pool Resources.* University of Michigan Press, Ann Arbor.
Ostrom, E., Burger, J., Field, C.B., Norgaard, R., and Policansky, D., 1999. Revisiting the commons: local lessons, global challenges. *Science,* 284: 278–282.
Pearce, D.W. and Atkinson, G.D., 1993. Capital theory and the measurement of sustainable development: an indicator of "weak" sustainability. *Ecol. Econ.,* 8: 103–108.
Peterjohn, W.T., and Correll, D.L., 1984. Nutrient dynamics in an agricultural watershed: observations on the role of a riparian forest. *Ecology,* 65: 1466–1475.
Peterson, S., 1994. Software for model building and simulation: an illustration of design philosophy. In *Modeling for Learning Organizations,* J.D.W. Morecroft, and J.D. Sterman, Eds., Productivity Press, Portland, OR, pp. 291–300.
Pezzey, J., 1989. Economic analysis of sustainable growth and sustainable development. Environment Department Working Paper No. 15. The World Bank, Washington, D.C.
Phillips, L.D., 1990. Decision analysis for group decision support. In *Tackling Strategic Problems: The Role of Group Decision Support,* C. Eden and J. Radford, Eds., Sage Publishers, London.
Pinker, S., 1995. *The Language Instinct. How the Mind Creates Language.* Harper Perennial Books, New York.
Rao, S.S., 1994. Welcome to open space. *Training,* April: 52–55.
Rastetter, E.B., King, A.W., Cosby, B.J., Hornberger, G.M., O'Neill, R.V., and Hobbie, J.E., 1992. Aggregating fine-scale ecological knowledge to model coarser-scale attributes of ecosystems. *Ecol. Appl.,* 2: 55–70.
Reich, P.B., Walters, M.B., and Ellsworth, D.S., 1992. Leaf Life-Span in Relation to Leaf, Plant, and Stand Characteristics among Diverse Ecosystems. *Ecol. Monogr.,* 62: 365–392.
Reuveny, R., and Maxwell, J.W., 1999. Continuing Conflict, Population, and Renewable Resources. School of Public and Environmental Affairs, Working Paper, Bloomington, IN.
Richmond, B., and Peterson, S., 1994. *STELLA II Documentation.* High Performance Systems, Inc., Hanover, NH.
Roberts, E.B., 1978. *Managerial Applications of System Dynamics.* Productivity Press, Portland, OR.
Roberts, H.H., Adams, R.D., and Cunningham, R.H., 1980. Evolution of the sand-dominated subaerial phase, Atchafalaya delta, Louisiana. *Am. Assoc. Petrol. Geol. Bull.,* 64: 264–279.
Robinson, J.B., 1991. Modeling the interactions between human and natural systems. *Int. Soc. Sci. J.,* 130: 629–647.
Robinson, J.B., 1992. Of maps and territories: the use and abuse of socioeconomic modeling in support of decision-making. *Technol. Forecasting Soc. Change,* 42: 147–164.

Rosenhead, J., Ed., 1989. *Rational Analysis of a Problematic World.* John Wiley & Sons, Chichester, England.
Ross, M., 1987. Industrial energy conservation and the steel industry of the United States. *Energy,* 12: 1135–1152.
Ruth, M., 1995. Technology change in US iron and steel production: implications for material and energy use, and CO_2 emissions. *Res. Pol.,* 21: 199–214.
Ruth, M., 1997. *Energy Use and CO_2 Emissions in a Dematerializing Economy: Examples from Five U.S. Metals Sectors.* Mimeograph. Center for Energy and Environmental Studies, Boston University, Boston, MA.
Ruth, M., and Dell'Anno, P., 1997. An industrial ecology of the U.S. glass industry. *Res. Pol.,* 23: 109–124.
Ruth, M., and Harrington, T., 1997. Dynamic of material and energy use in US pulp and paper manufacturing. *J. Ind. Ecol.,* 1: 147–168.
Rykiel, E.J., 1996. Testing ecological models: the meaning of validation. *Ecol. Model.,* 90: 229–244.
Sanchiro, J.N., and Wilen, J.N., 1999. Bioeconomics of spatial exploitation in a patchy environment, *JEEM,* in press.
Scaife, W.W., Turner, R.E., and Costanza, R., 1983. Coastal Louisiana recent land loss and canal impacts. *Environ. Manage.,* 7: 433–442.
Senge, P.M., 1990. *The Fifth Discipline.* Doubleday, New York.
Senge, P.M., and Sterman, J.D., 1994. Systems thinking and organizational learning: acting locally and thinking globally in the organization of the future. In *Modeling for Learning Organizations,* J.D.W. Morecroft and J.D. Sterman, Eds., Productivity Press, Portland, OR, pp. 195–216.
Sengupta, N., 1991. *Managing Common Property: Irrigation in India and the Philippines.* Sage Publication, New Delhi.
Sengupta, N., and Sampath, R.K., 1995. On a composite measure of performance for irrigation systems. *Int. Comm. Irrig. Drain. J.,* 44: 11–20.
Shabman, L., 1988. Reducing nitrogen pollution from crop production systems. Chesapeake Research Consortium. *Fourth Annual Conference on the Economics of Chesapeake Bay Management,* Baltimore MD. May 10, 1988, conference paper.
Simon, H.A., 1956. *Administrative Behavior.* Wiley & Sons, New York.
Simon, H.A., 1979. Rational decision-making in business organizations. *Am. Econ. Rev.,* 69: 493–513.
Simon, C.P., and Blume, L., 1994. *Mathematics for Economists.* W. W. Norton, New York.
Simpson, T., 1996. Maryland Department of Agriculture, personal communication.
Sinclair, M., 1988. *Marine Populations: an Essay on Population Regulation and Speciation.* Washington Sea Grant Program, University of Washington Press.
Sklar, F.H., Costanza, R., and Day, J.W., Jr., 1985. Dynamic spatial simulation modeling of coastal wetland habitat succession. *Ecol. Model.,* 29: 261–281.
Sklar, F.H., White, M.L., and Costanza, R., 1989. The Coastal Ecological Landscape Spatial Simulation (CELSS) model: structure and results for the Atchafalaya/Terrebonne study area. U. S. Fish and Wildlife Service, Division of Biological Services, Washington, D.C.
Small, L.E., and Svendse, N., 1992. A framework for assessing irrigation performance. Working Papers on Irrigation Performance, International Food Policy Research Institute, Washington, D.C., August.
Solow, R.M, 1956. A contribution to the theory of economic growth. *Q. J. Econ.,* 70: 65–94.

Sonnenschein, H., 1974. Market excess demand functions. *Econometrica*, 40: 549–563.
Sparrow, F.T., 1983. *Energy and Material Flows in the Iron and Steel Industry.* Argonne National Laboratory, ANL/CNSV-41, Argonne, IL.
Steinhilber, P., 1996. Coordinator of the Nutrient Management Program (NMP), personal communication.
Tang, S.Y., 1992. *Institutions and Collective Action: Self-Governance in Irrigation.* ICS Press, San Francisco.
Taub, F.B., 1987. Indicators of change in natural and human impacted ecosystems: Status. In *Preserving Ecological Systems: The Agenda for Long-Term Research and Development,* S. Draggan, J.J. Cohrssen, and R.E. Morrison, Eds., Praeger, New York, pp. 115–144.
Taub, F.B., 1989. Standardized aquatic microcosm—development and testing. In *Aquatic Ecotoxicology: Fundamental Concepts and Methodologies.* Vol. II, A. Boudou and F. Ribeyre, Eds., CRC Press, Boca Raton, FL., pp. 47–92.
Turner, M.G., 1989. Landscape ecology: the effect of pattern on process. *Annu. Rev. Ecol. Syst.,* 20: 171–197.
USACERL, 1993. *GRASS Version 4.1. User's Reference Manual.* Open GRASS Foundation Center for Remote Sensing, Boston University.
Van Heerden, I.L., and Roberts, H.H., 1980a. The Atchafalaya delta—Louisiana's new prograding coast. *Trans. Gulf Coast Assoc. Geol. Soc.,* 30: 497–506.
Van Heerden, I.L., and Roberts, H.H., 1980b. The Atchafalaya delta—rapid progradation along a traditionally retreating coast (South Central Louisiana). *Z. Geomorph. N.F.,* 34: 188–201.
Vennix, J.A.M., and Gubbels, J.W., 1994. Knowledge elicitation in conceptual model building: a case study in modeling a regional Dutch health care system. In *Modeling for Learning Organizations,* J.D.W. Morecroft, and J.D. Sterman, Eds., Productivity Press, Portland, OR, pp. 121–146.
Villa, F., 1997. Usage of the Model Performance Evaluation Software. IEE Internal Report, University of Maryland.
Villa, F., and Costanza, R., (in preparation). Performance and goodness of fit of complex simulation models: a multi-criteria approach to model calibration using the Model Performance Index (MPI) framework. In *Spatially Explicit Landscape Simulation Modeling,* R. Costanza and A. Voinov, Eds., Springer, New York.
VIMS SAV Mapping Lab, 1997. Historical SAV distribution (1971–1995): *Totals by Chesapeake Bay Program Segment,* http://www.vims.edu/bio/sav/index.html.
Voinov, A.A., Fitz, H.C., and Costanza, R., 1998. Surface water flow in landscape models: 1. Everglades case study. *Ecol. Model.,* 108: 131–144
Voinov, A., Voinov, H., and Costanza, R., 1999a. Surface water flow in landscape models: 2. Patuxent watershed case study. *Ecol. Model.,* 119: 211–221.
Voinov, A., Costanza, R., Wainger, L., Boumans, R., Villa, F., Maxwell, T., and Voinov, H., 1999b. Patuxent landscape model: integrated ecological economic modeling of a watershed. *J. Ecosyst. Model. Software,* 14: 473–491.
Walters, C.J., 1986. *Adaptive Management of Renewable Resources.* McGraw-Hill, New York.
Weissing, F., and Ostrom, E., 1991. Irrigation institutions and the games irrigators play: rule enforcement without guards. In *Game Equilibrium Models II: Methods, Morals, and Markets,* R. Selten, Ed., Springer-Verlag, Berlin, pp. 188–262.
Weissing, F., and Ostrom, E., 1993. Irrigation institutions and the games irrigators play: rule enforcement on government- and farmer-managed systems. In *Games*

in Hierarchies and Networks: Analytical and Empirical Approaches to the Study of Governance Institutions, F.W. Scharpf, Ed., Westview Press, Boulder, CO., pp. 387–428.

Westenholme, E.F., 1990. *System Inquiry: A System Dynamics Approach.* John Wiley & Sons, Chichester, England.

Westenholme, E.F., 1994. A systematic approach to model creation. In *Modeling for Learning Organizations,* J.D.W. Morecroft and J.D. Sterman, Eds., Productivity Press, Portland, OR, pp. 175–194.

Weston, R.F., and Ruth, M., 1997. A dynamic hierarchical approach to understanding and managing natural economic systems. *Ecol. Econ.,* 21: 1–17.

Wilson, J.A., 1990. Fishing for knowledge. *Land Econ.,* 66: 12–29.

Wilson, J.A., 1997. Maine fisheries management initiative. In *The Social Impacts of Individual Transferable Quotas,* G. Palsson, Ed., TemaNord, Copenhagen, pp. 335–353.

Wilson, J.A., French, J., Kleban, P., McKay, S., and Townsend, R., 1991. Chaotic dynamics in a multiple species fishery: a model of community predation. *Ecol. Model.,* 58: 303–322.

Wilson, J., Kleban, P., Metcalfe, M., and Acheson, J., 1994. Chaos, complexity and community management of fisheries. *Mar. Pol.,* 18(4): 291–305.

Wilson, J.A., Kornfield, I., Hunt von Herbing, I., Tupper, M., 1998. *Implications of Localized Fishery Stocks.* NRAES Press, Ithaca, NY.

World Commission on Environment and Development, 1987. *Our Common Future.* Oxford University Press, Oxford.

Wroblewski, J.S. and Hofmann, E.E., 1989. U.S. interdisciplinary modeling studies of coastal-offshore exchange processes: past and future. *Prog. Oceanogr.,* 23: 65–99.

Wulff, F., and Ulanowicz, R.E., 1989. A comparative anatomy of the Baltic Sea and Chesapeake Bay ecosystems. Chapter 11. In *Network Analysis of Marine Ecology: Methods and Applications,* F. Wulff, J.G. Field, and K.H. Mann, Eds., Coastal and Estuarine Studies Series, Springer-Verlag, Heidelberg, pp. 232–256.

Zeeman, M.L., 1993. Hopf bifurcations in competitive three-dimensional Lotka-Volterra systems. *Dynam. Stab. Syst.,* 8: 189–217.

Zucker, L.G., 1986. Production of trust: institutional sources of economic structure. *Res. Org. Struct.,* 8: 53–111.

Index

D

E

F

G

H

M

R

S

T

U

V

W

For Product Safety Concerns and Information please contact our EU
representative GPSR@taylorandfrancis.com
Taylor & Francis Verlag GmbH, Kaufingerstraße 24, 80331 München, Germany

www.ingramcontent.com/pod-product-compliance
Ingram Content Group UK Ltd.
Pitfield, Milton Keynes, MK11 3LW, UK
UKHW010956180926
475701UK00004B/404

* 9 7 8 0 3 6 7 3 9 7 7 5 3 *